W9-BHU-147

SPACE REMOTE SENSING SYSTEMS

An Introduction

SPACE REMOTE SENSING SYSTEMS

An Introduction

H. S. Chen

ROCKWELL INTERNATIONAL
DOWNEY, CALIFORNIA
AND
UNIVERSITY EXTENSION
UNIVERSITY OF CALIFORNIA, LOS ANGELES
LOS ANGELES, CALIFORNIA

1985

ACADEMIC PRESS, INC.
Harcourt Brace Jovanovich, Publishers
Orlando San Diego New York Austin
London Montreal Sydney Tokyo Toronto

ACADEMIC PRESS, INC.
Orlando, Florida 32887

United Kingdom Edition published by
ACADEMIC PRESS INC. (LONDON) LTD.
24–28 Oval Road, London NW1 7DX

Library of Congress Cataloging in Publication Data

Chen, H. S. (Hsi Shu)
Space remote sensing systems.

Includes index.
1. Remote sensing–Equipment and supplies. I. Title.
G70.6.C47 1985 621.36'78 85-1375
ISBN 0–12–170880–2 (alk. paper)
ISBN 0–12–170881–0 (paperback)

PRINTED IN THE UNITED STATES OF AMERICA

87 88 9 8 7 6 5 4 3 2

To my family

Contents

Chapter 8 Active Synthetic Aperture Radar and Scatterometer Systems

Chapter 9 Low-Earth-Orbit Large Satellite Systems

Chapter 10 Geosynchronous-Earth-Orbit Large Satellite Systems

Preface

The space remote sensing system is a modern high-technology field that has been developed from earth sciences, engineering, and space systems technology for environmental protection, resource monitoring, climate prediction, weather forecasting, ocean measurement, and many other applications.

This book is written for scientists, engineers, and seniors or graduate students who are interested in the field of space remote sensing systems, especially for the introduction of space sensor systems and their applications.

In the field of space remote sensing, the most important recent advances have probably been the operation of the Space Shuttle for science and application missions and the development of the Space Station system for the benefit of us all. We are now at the beginning of a new era in space remote sensing associated with new space systems. The exploitation of new missions and new sensors, in the form of new technology, may enable the current techniques to be extended to better and more useful space programs. Such progress may well have a profound impact on future technology, for instance, not only on the Space Shuttle program but also the Space Station.

Most of the contents of this book are based on classroom material that the author taught in short courses in 1982, 1983, and 1984 under the title Advanced Space Sensor Systems. This book provides definitions of new concept and technology development of space remote sensing systems applicable to the 1990s and beyond. Based on the state of the art, this book introduces future technological approaches that should allow the reader to gain knowledge in the area of remote sensing of the earth and other special targets.

In this book, an in-depth presentation of state-of-the-art space remote sensing systems is made. Ten broad areas are covered: Chapter 1 describes the science of the atmosphere and the earth's surface. Chapter 2 discusses spaceborne radiation collector systems. Chapter 3 discusses space detector and CCD systems. Chapter 4 presents passive space optical radiometer systems. Chapter 5 discusses passive space optical spectrometer systems. Chapter 6 describes passive space microwave radiometer systems. Chapter 7 discusses active space lidar systems. Chapter 8 presents active space synthetic aperture radar and scatterometer systems. Chapter 9 discusses low-earth-orbit large satellite systems and applications. Chapter 10 describes the geosynchronous-orbit large satellite systems and applications. In each chapter, advanced concepts and technologies are presented and reviewed for future systems and applications.

The author has been involved over the years in research and analysis of space sensor systems with many professionals in industry, government, and universities. To attempt to list all their names would be inappropriate here. Nevertheless, their contributions are much appreciated, for without them this book would not have been possible.

The contents of this book are the author's own presentation, which does not represent the policy of, or endorsement from or for, any company or institute.

In selecting references for inclusion in this book, the goal has not been to present a very long and complete set but to indicate to the reader the more important papers and also some key references that can be used to collect a fairly complete set for further study.

Special thanks go to the author's family for the cooperation that was given during the writing of this book. Finally, the author would like to thank the staff of Academic Press for their courtesy and assistance during the preparation of the manuscript.

The author is now affiliated with Hughes Aircraft Company, El Segundo, California.

Chapter 1 Earth's Atmosphere and Its Surface Systems

1.1 Structure of the Atmosphere

The earth's atmospheric air is a mixture of various constituents, the principal ones being nitrogen, oxygen, argon, carbon dioxide, and water vapor. Suspended solid and liquid particles of natural and man-made origin are always present in the atmosphere; they are called atmospheric aerosol.

The atmosphere is not homogeneous throughout. Its state and properties vary significantly with altitude as well as from point to point and time to time. The atmosphere may be divided vertically into a number of layers differing in composition, temperature, and other physical properties.

The troposphere is the lowest layer of the atmosphere. The upper limit of the troposphere lies at a height ranging from 7 to 18 km, depending on latitude, time of year, and properties of the earth's surface. The most characteristic feature of the troposphere is a temperature that falls off with altitude. The troposphere is a region of particularly strong vertical motion and heat exchange with the earth's surface.

The layer above the troposphere is the stratosphere, at a height from 11 to 50 km. The stratosphere is distinguished by the fact that its temperature remains nearly constant or increases with altitude. Starting from a height of about 35 km, the temperature rises considerably, and this temperature rise is due to the absorption of solar radiation in the ozone layer that is found at these altitudes. Above 55 km lies the mesosphere, characterized by a temperature that falls off with altitude, which is followed by the transition from the mesosphere to the next layer, called the thermosphere. The thermosphere is the thickest layer of all, characterized by a

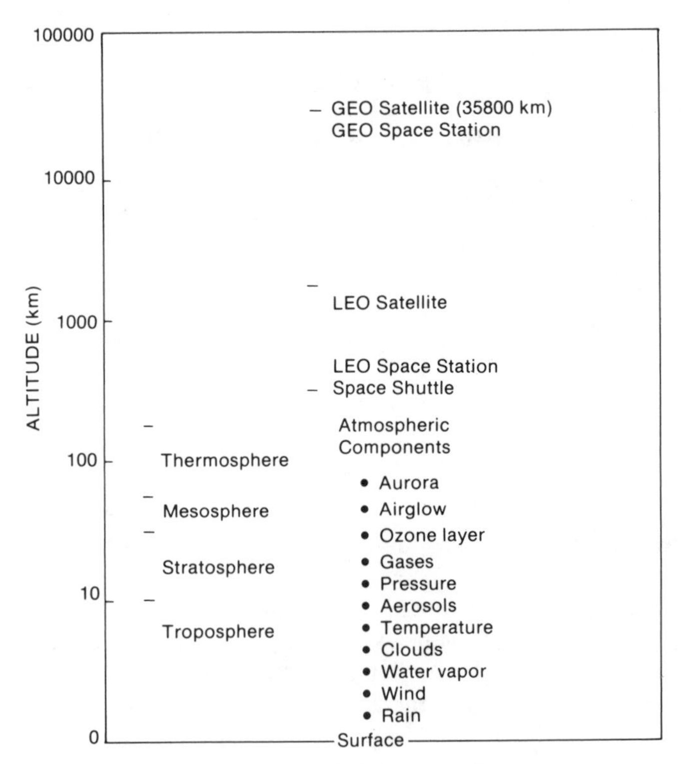

Fig. 1.1. Structure of the atmosphere.

continuous rise in temperature with altitude up to its upper boundary, which occurs at a height of several hundred kilometers. Figure 1.1 shows the vertical structure of the atmosphere.

The attenuation of solar and earth radiation in the atmosphere is the consequence of the atmosphere's absorption and scattering. The principal absorbers of radiation in the atmosphere are water vapor, carbon dioxide, ozone, oxygen, and aerosol. Water vapor is the most important factor in the absorption of radiation in the atmosphere, not only because it is present in large amounts, but also because of the very great number of lines and bands in its complex absorption spectrum.

The water vapor absorption bands are located at 0.72, 0.82, 0.93, 1.13, 1.38, 1.86, 2.01, 2.05, 2.68, 3.6, 4.5, 6.3, and 17 μm and at microwave 0.16 and 1.5 cm. Water vapor continuum absorption is also important for space remote sensing applications.

Carbon dioxide has a number of absorption bands in the infrared region of the spectrum. The most important band is the broad one centered at about 14.7 μm, where the maximum thermal emission of the atmosphere occurs. Carbon dioxide has a series of other bands centered at about 1.4, 1.6, 2.0, 2.7, 4.3, 4.8, and 5.2 μm.

Ozone, which occurs in a layer extending from 10 to 60 km in height and is concentrated at about 22 km, has extremely high absorption in the ultraviolet region. The principal absorption bands of ozone in the UV, visible, and IR are centered at about 0.15, 0.25, 0.30, 0.60, and 9.6 μm.

Oxygen has absorption in the visible, UV, and microwave regions. The visible spectrum is centered between about 0.69 and 0.76 μm. The UV absorption spectrum is located in the region between 0.18 and 0.25 μm, and the microwave spectrum is centered at about 0.27 and 0.5 cm.

Airglow and aurora constitute the atmospheric radiation emitted by the upper atmosphere in the ultraviolet, visible, and near-infrared parts of the spectrum. Airglow is present at all times and places, and aurora occurs predominantly at high latitudes, less frequently at middle latitudes, and only occasionally at low latitudes. Aurora is typically much brighter than airglow. Both airglow and aurora can be used as special emitted radiation to determine the temperature and wind field of the upper atmosphere from space spectrometer measurements. The oxygen green line at 5577 Å, the oxygen red lines at 6300 and 6364 Å, the atomic oxygen line at 7319 Å, and the OH lines at 6830 and 6876 Å are the emitted radiation used for space remote sensing applications. The emitted radiation heights vary from 85 to 250 km.

The earth's clouds form as a result of the condensation of water vapor contained in the atmosphere. Half of the earth's area is covered by clouds. There are three different types of clouds: water, ice, and mixed water and ice. Clouds can be categorized according to their height above the ground as low clouds (up to 2.5 km), middle clouds (2.5–6.0 km), and high clouds (6.0 km and above). The observation of clouds from space determines their type, altitude, velocity, particle density, and size.

1.2 Transmittance and Radiance of the Atmosphere

Radiance from background and from targets in the earth's atmosphere and on its surface is attenuated in its passage through the atmosphere. There has been a great deal of research from which methods have been derived for the calculation of atmospheric transmittance and radiance. Two general categories of such calculation methods, namely the line-by-line fast atmospheric signature code (FASCOD) method and the empirical low resolution transmittance (LOWTRAN) method, are perhaps the most widely known and used.

The line-by-line method employs a set of parameters that describe the molecular lines in which radiation is absorbed and emitted for high-resolution spectral application. The only complete set of line data currently in existence is that compiled by the Air Force Geophysics Laboratory (AFGL). The compilation contains the following line parameters: line center, line strength, line half-width, and energy of the lower state. Most of the line-by-line techniques developed are designed to minimize the computation cost. The best FASCOD line-by-line method is designed to achieve as high an accuracy as possible without overburdening the computation with large costs and long computer times. Even so, the cost of performing a direct integration over the individual lines is far greater than that of using band models.

The absorption coefficient for a pressure-broadened (Lorentz) line is given by the Lorentz function,

$$A(\nu) = S\alpha/[\pi(\nu - \nu_0)^2 + \alpha^2] \tag{1.1}$$

where $A(\nu)$ is the line absorption coefficient, ν the wave number, ν_0 the line-center wave number, S the line intensity, and α the pressure-broadened half-width.

The variation of the pressure-broadened half-width with pressure and temperature is given by

$$\alpha(P, T) = \alpha_0(P_0, T_0)(P/P_0)(T_0/T)^{1/2} \tag{1.2}$$

where α_0 is the pressure-broadened half-width at the earth's surface, T_0 the temperature at the earth's surface, and P_0 the pressure at the earth's surface. The half-widths in the AFGL line compilation are typically of the order of 0.08 cm^{-1}.

The absorption coefficient for a line broadened by thermal motion is given by the Gaussian function in terms of the Doppler line width as

$$\alpha_D = (\nu_0/c)[2(\ln 2)kT/(M/N_0)]^{1/2} \tag{1.3}$$

where α_D is the Doppler half-width, T the line temperature, c the velocity of light, k the Boltzmann constant, M the molecular weight of the molecule type, and N_0 Avogadro's number.

The LOWTRAN method has been revised many times and has also received a wide distribution; it is easy and inexpensive to use and is reasonably accurate. The LOWTRAN method is strictly empirical, deriving its functional form from a graphical fit to molecular line data obtained in experimental investigations. In the LOWTRAN method, the average transmittance spectral resolution is 20 cm^{-1}.

The LOWTRAN program has the option of calculating atmospheric and earth radiance. A numerical evaluation of the integral form of the equation of radiative transfer is used in the program. Local thermodynamic equilibrium is assumed in the atmosphere.

Blackbody radiance emitted by the earth and the atmosphere, solar radiance scattered by the atmosphere, solar radiance reflected from the surface, and total transmittance with aerosols and cirrus cloud effects are all included in the LOWTRAN computer code. The spectral radiance from a given atmospheric path length is calculated by dividing the atmosphere into a series of isothermal layers and summing the radiance contributions from the individual layers along the space sensor field of view.

In general, fairly good agreement has been found between space radiance measurements and LOWTRAN calculations. [The LOWTRAN and FASCOD codes are available from the National Climate Center, Federal Building, Asheville, North Carolina 28801.]

1.3 Surface Reflection and Emission

It is well known that naturally occurring surfaces have widely varying reflection properties, the variation being in terms of the dependence of reflectance on wavelength, angle of incidence, angle at which the surface is viewed from space, and the physical characteristics of the surface itself.

Two quite distinct processes are responsible for the reflection of radiation at the earth's surface. The first is specular (mirrorlike) reflection. The second is diffuse reflection (i.e., from a rough surface).

It can be shown, using Snell's law of refraction and standard trigonometric identities, that the specular reflectance can be written (one of Fresnel's equations)

$$R = \frac{1}{2}\left[\frac{\sin^2(i-j)}{\sin^2(i+j)} + \frac{\tan^2(i-j)}{\tan^2(i+j)}\right] \tag{1.4}$$

where i is the angle of incidence and j the angle of refraction.

When the angle of incidence satisfies the condition

$$\tan\theta_B = n_2/n_1 \tag{1.5}$$

then θ_B is called Brewster's angle, or the polarizing angle.

The Fresnel reflectivity for normal incidence is

$$R = [(n_2 - n_1)/(n_2 + n_1)]^2 \tag{1.6}$$

Putting $n_2 = 1.3$ in this equation shows that the reflectivity of a typical air–water interface is about 1.69% for normal incidence.

No general theory is completely valid for the reflection from the earth's natural surfaces, although all formulas for diffuse reflection incorporate Lambert's law in one form or another. Lambert's law is phenomenologically formulated from the fact that a rough surface irradiated with constant intensity appears uniformly light at all angles of observation. The diffuse reflected radiation is taken to be unpolarized, regardless of the state of polarization of the incident radiation.

The real reflection from the earth's surface is composed of both diffuse and specular reflection, the relative proportion of each depending on the nature of the reflecting medium.

Glitter from the sea is a phenomenon that bears upon reflection of solar radiation. The glitter arises when the sea surface is roughened by wind and the image of the sun formed by specular reflection explodes into glittering points. This occurs because seawater facets are oriented in such a way that they reflect sunlight to the space sensor. Increasing roughness will enlarge the width of the glittering band. The sun's glitter image is most spectacular at solar elevations of 30 to 35°. The glitter pattern becomes narrower as the sun sets. The polarization of the reflected glitter is dependent on both solar elevation and the direction of reflection. The polarization at the maximum radiance is near zero at $i = 0°$ and increases with zenith distance to 100% at $i = 30–45°$, subsequently decreasing to 20% at a large angle of incidence.

The solar radiation that penetrates to the earth's surface is partially reflected by the surface. Reflection also takes place in the atmosphere, mainly from clouds. Reflectance is the ratio of the radiant intensity reflected by a given surface to the radiant intensity incident on this surface. The reflectance is usually expressed in percentages for specular reflection. For diffuse reflection, albedo is used; this is the ratio of the reflected radiant flux to the flux of incident radiation. Surface albedo is also expressed in percentages for remote sensing of the earth. Table 1.1 lists albedo values for different natural surfaces.

The emissivity of the earth's surface is important in determining the surface temperature. The emissivity depends on the wavelength and on the nature of the surface. The 10–12-μm band is the most suitable one for the determination of surface temperatures by space remote sensing, since it is situated in the atmospheric window and the relative emissivity of surfaces in this interval is comparatively stable and close to unity. Table 1.2 presents a summary of measurements of emissivity in the atmospheric window of 10 to 12 μm.

Radiation emitted from the earth's surface can be considered blackbody radiation and is a function of the surface temperature and the wavelength. Planck's equation describes the spectral distribution of the radia-

Table 1.1

Different Natural Surface Albedos

Surface	Albedo (%)
Snow	60–90
Water	1–9
Soil	8–30
Sand	20–40
Vegetation	5–30
Ice	25–40
Cloud	30–90
Desert	20–35

tion from a blackbody as

$$E = \frac{2\pi hc^2}{\lambda^5} \frac{1}{\exp(ch/kT\lambda) - 1} \tag{1.7}$$

where E is spectral irradiance in watts per square centimeter micrometer, λ wavelength in micrometers, h Planck's constant (6.6256×10^{-34} W sec^2), T absolute temperature in Kelvins, c velocity of light (2.998×10^{10} cm/sec), and k Boltzmann's constant (1.380×10^{-23} W sec/K).

From the inversion of the radiative transfer equation, the earth's surface temperature can be determined by space remote sensing; also, the atmospheric temperature and species can be determined from measurements of the emission from the atmosphere.

Figure 1.2 shows blackbody emission as a function of wavelength for several temperatures.

Every space remote sensing system employs certain spectral bands for earth and atmospheric observations. In the electromagnetic spectral domain each wavelength region has been named for better definition. Table

Table 1.2

Surface Emissivities at 10 to 12 μm

Surface	Emissivity	Surface	Emissivity
Water	0.993–0.998	Sand	0.949–0.962
Ice	0.98	Snow	0.969–0.997
Green grass	0.975–0.986	Granite	0.898
Sandy loam	0.954–0.968	Peat	0.979–0.983

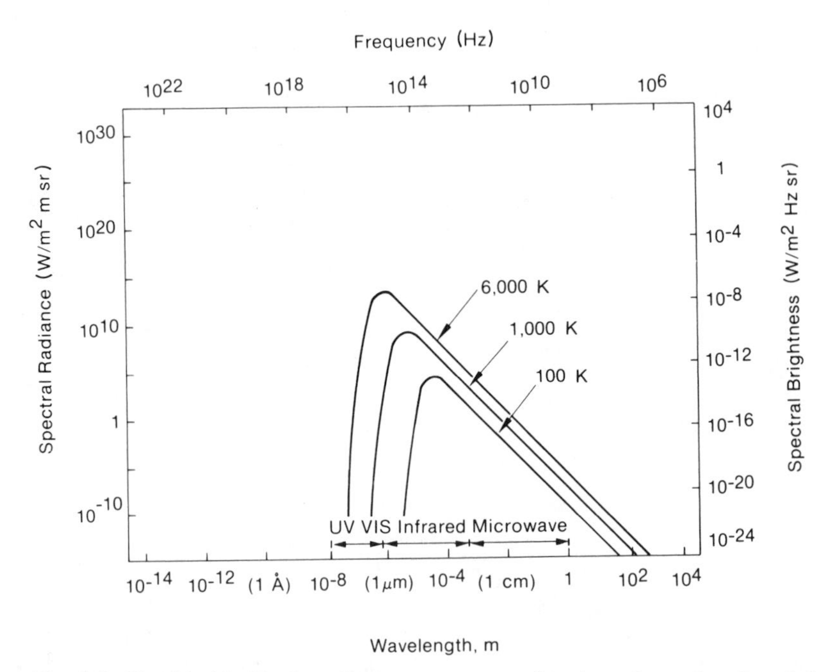

Fig. 1.2. Planck's blackbody radiation curves as a function of wavelength and frequency.

1.3 shows the names of the wavelength ranges used for remote sensing applications.

1.4 Solar Radiation

The sun, the star closest to us, is a yellow dwarf. Its mean radius is 695,500 km, or 109 times the radius of the earth. The distance of the earth from the sun changes over the year because of the ellipticity of the earth's orbit. The mean distance of the earth from the sun is 149.5 Mkm. Despite the enormous linear dimensions of the sun, the apparent angular size of the solar disk is small because of its great distance; its mean angular diameter is 31′59″.

The sources of solar energy flux are thermonuclear reactions taking place in the interior of the sun. The main solar emission flux is produced by the photosphere, which is the visible surface of the sun and has a temperature of about 6000 K. The photosphere is actually a very thin layer, having a thickness of only a few hundred kilometers.

Table 1.3

Names of Wavelength Regions

Spectrum name	Spectrum subname	Wavelength
Ultraviolet	Far ultraviolet (FUV)	0.01–0.20 μm
	Middle ultraviolet (MUV)	0.20–0.30 μm
	Near ultraviolet (NUV)	0.30–0.38 μm
Visible	Violet	0.38–0.45 μm
	Blue	0.45–0.49 μm
	Green	0.49–0.56 μm
	Yellow	0.56–0.59 μm
	Orange	0.59–0.63 μm
	Red	0.63–0.76 μm
Infrared	Near infrared (NIR)	0.80–1.50 μm
	Short-wavelength infrared (SWIR)	1.50–3.00 μm
	Middle-wavelength infrared (MWIR)	3.00–5.00 μm
	Long-wavelength infrared (LWIR)	5.00–15.0 μm
	Far infrared (FIR)	15.0–300.0 μm
Microwave	Submillimeter	0.01–0.10 cm
	Millimeter	0.10–1.00 cm
	Microwave	1.0–100.0 cm

Solar rays arriving from the limbs of the solar disk must traverse relatively thick layers of matter, so that there is a comparatively low emission flux from the deep, hot photospheric layers. This is the reason for the darkening of the limb of the solar disk. Moreover, the part of the radiation at relatively short wavelengths penetrates thick layers of matter with much more difficulty than does longwave radiation, so that in addition to the limb darkening there is a reddening of the emission from the limb. Consequently, the blackbody temperature closest to the temperature of the emission from limbs is somewhat lower, and the mean blackbody temperature obtained by integrating the solar emission flux over the entire disk is about 5750 K.

The solar constant is the magnitude of the flux of radiant energy from the sun per unit time across a unit surface onto a surface situated normal to the rays outside the earth's atmosphere at the earth's mean distance from the sun.

Since the earth's distance from the sun changes as a function of time, the solar constant I' outside the atmosphere at a time when the earth's distance from the sun is R will be different from I_0 and is given by

$$I_0 = (R^2/R_0^2)I' \tag{1.8}$$

where R_0 is the mean distance of the earth from the sun. The variation of the solar constant is a function of time. It is a maximum when the earth is at the beginning of January and a minimum at the beginning of July. The variation through the year amounts to $\pm 3.4\%$ of its mean value.

It is evident from the definition of the solar constant that the mean value of the energy emitted by the sun per minute is given by

$$E = 4\pi R_0^2 I_0 \tag{1.9}$$

The energy emitted by the sun can be calculated from this expression and turns out to be 3.85×10^{23} kW. The part of this energy intercepted by the earth is given by

$$E = \pi r_0^2 I_0 \tag{1.10}$$

where r_0 is the mean radius of the earth. The calculated energy is 1.79×10^{14} kW.

Solar radiation can be used in solar occultation techniques for measurements of the earth's atmospheric temperature and species vertical profiles. Backscattered solar radiation can be used to determine the vertical profile of ozone. Solar radiation reflected from the earth's surfaces has been used for earth resources observations from space. The earth radiation budget satellite measures the solar radiation constantly in space. Figure 1.3 shows the solar radiation as a function of wavelength.

1.5 Imaging through the Atmosphere

There are passive and active methods for space remote imaging of the earth's surfaces and clouds. Reflected solar radiance and the earth's thermal emission radiance are the major sources for passive space sensor system applications. Reflected laser beams and microwaves are the major sources for active applications. Reflected and scattered solar radiance can be used only in the daytime for earth imaging; thermal emission from the earth can be used day and night. Of course, the active imaging systems can be used at any time for detection at different looking angles. The most useful infrared space imaging system is the geosynchronous operational environment satellite (GOES) system space sensor, which provides hourly environmental picture analysis. The most useful visible and near-infrared space imaging system is the Landsat-type space sensor, which provides monthly, seasonal, or yearly analysis of earth resources. The

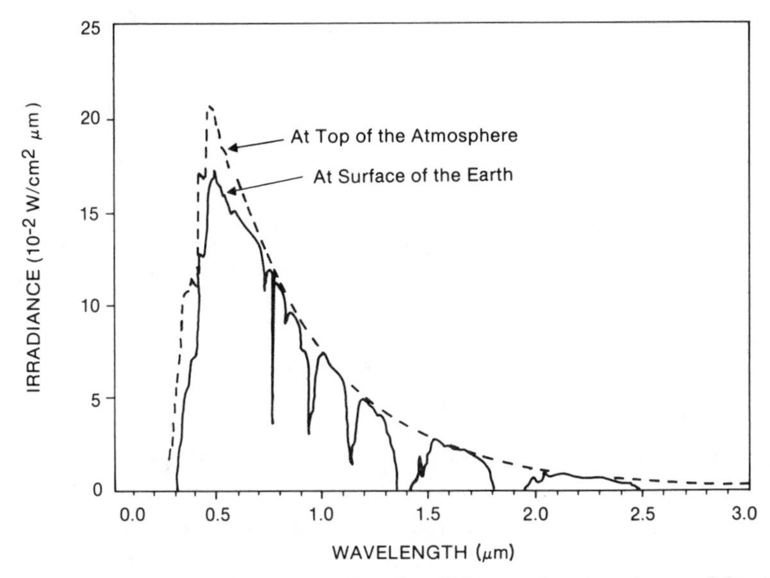

Fig. 1.3. Solar spectral radiance as a function of the wavelength at the top of the atmosphere and at the surface of the earth.

most useful microwave space sensor is the synthetic aperture radar (SAR) system for surface imaging applications. The new synthetic aperture lidar system will be ready for space applications in the 1990s. Table 1.4 shows the ground resolution expected for each space imaging system for the 1990s.

The electromagnetic radiation formed within the earth's atmosphere originates from earth surface emission, atmospheric emission, scattering of solar radiation within the atmosphere, reflection of solar radiation from the earth's surface, and other special emission sources. The radiative transfer equation for earth and atmospheric radiation can be expressed as follows under assumptions of plane-parallel atmosphere, local thermody-

Table 1.4

Ground Resolution of Space Imaging Systems

Space sensor system	Ground resolution	Source
Passive visible	1–30 m	Reflection (daytime only)
Passive infrared	5–120 m	Emission (day and night)
Passive microwave	1–15 km	Emission (day and night)
Active synthetic aperture radar	2–30 m	Reflection (not for cloud)
Active synthetic aperture lidar	0.1–1 m	Reflection (day and night)

namic equilibrium, single scattering, and Lambertian surface reflection (LOWTRAN approach):

$$L(\lambda) = B_{SS}\tau + \int B_L\, d\tau + (1 - \varepsilon) \int B_L\, d\tau + I_0 \int \tau \left[P_A \frac{d\tau_A}{\tau_A} + P_M \frac{d\tau_M}{\tau_M} \right] + I_0 \tau a \frac{\cos\theta}{2\pi} \tag{1.11}$$

where B is Planck's function, B_{SS} the earth surface emission, τ the transmittance, B_L the atmospheric emission of each layer, I_0 solar radiation, P_A the aerosol scattering phase function, τ_A the aerosol transmittance, P_M the molecular single scattering function, λ the wavelength, τ_M the molecular transmittance, θ the solar zenith angle, a the surface albedo, and ε the emissivity of the earth's surface.

In this equation the first term is the radiation emitted from the earth's surface to space; the second, the atmospheric radiation emitted to space; the third, the emitted atmospheric radiation reflected by the surface to space; the fourth, the solar radiation scattered upward within the atmosphere by the aerosol and molecular species; and the fifth, the solar radiation reflected from the earth's surface or a special target to space.

In visible and near-infrared daytime imaging applications, the radiative transfer equation can be expressed as

$$L(\lambda) = I_0 \int \tau \left[P_A \frac{d\tau_A}{\tau_A} + P_M \frac{d\tau_M}{\tau_M} \right] + I_0 \tau a \frac{\cos\theta}{2\pi} \tag{1.12}$$

Since thermal imaging is one of the most useful space imaging techniques, we shall review the science of thermal remote sensing. The radiance leaving the earth–atmosphere system that can be sensed by a space sensor is the sum of the radiation emissions from the surface and each atmospheric layer transmitted to the top of the atmosphere and the aperture of the space sensor. If the earth's surface is assumed to be a blackbody emitter, the upwelling radiance $L(\nu)$ for a cloudless atmosphere is given by

$$L(\nu) = \varepsilon B(\nu, T_S)\tau(\nu, P_S) + \int_{\tau(\nu, P_S)}^{1} B(\nu, P)\, d\tau(\nu, P) \tag{1.13}$$

where L is the total earth and atmospheric radiance, B Planck's function, τ the transmittance, T the temperature, ν the frequency of the radiation, P the atmospheric pressure, and ε the emissivity of the earth's surface.

The first term is the thermal radiation emitted by the surface of the earth and transmitted by the overlying atmosphere. The second term is the thermal radiance emitted by the atmosphere in the upward direction. For surface imaging, the best spectral selection will be the second-term radiance near the minimum value. For atmospheric sounding, the first-term radiance should be at the minimum value for spectral band selection.

In the case of land, the surface temperature varies considerably with the type of surface and also with the time of day. In addition, the surface temperature can be considerably different from the temperature of the atmosphere near the surface.

Three spectral regions are available for imaging in the infrared when the atmosphere is transparent: 3.5–3.9 μm, 10.3–11.3 μm, and 11.5–12.5 μm. The main disadvantage of the first band is that some reflected solar radiation is present, so that accurate temperature measurement is possible only when sunlight is not present.

In order to determine temperature T it is necessary to know the emissivity ε of the surface, the transmittance τ of the atmosphere, and whether any clouds are present. For parts of the earth that are well covered with vegetation the emissivity is approximately unity in all spectral bands. The emissivity of the ocean surface is very close to unity throughout the window region.

The solar radiation reflected from the earth's surface in the visible and near-infrared has been used in space remote sensing for many years. The reflected solar radiation is a function of the surface albedo, the solar illumination angle at the earth's surface, and the transmittance of the atmosphere.

The potential for synthetic aperture radar in space remote sensing has been demonstrated in recent years. Synthetic aperture radar data are useful for geologic and ocean surface mapping. New synthetic aperture lidar may be extremely useful for certain applications in coming years.

1.6 Sounding of the Atmosphere

Atmospheric sounding in the earth's atmosphere results from radiative transfer inversion of oxygen, carbon dioxide, water vapor, and ozone. Radiance in bands of gases with a fixed concentration can be used to measure vertical temperature profiles. For gases with a variable concentration, space remote sounding can be used to estimate the vertical concentration profiles.

For a particular atmospheric absorption band to be used for temperature sounding the following conditions must be satisfied:

(a) The atmospheric emitting constituent should be uniformly mixed in the atmosphere so that the emitted radiation can be considered a function of the temperature distribution only.

(b) The atmospheric absorption band involved should not be overlapped by strong absorption bands of other constituents.

(c) Local thermodynamic equilibrium (LTE) should apply in the region of temperature sounding.

Consider first, in an infinitely deep atmosphere that is horizontally stratified, a slice at temperature T containing a path length dz of absorber in the vertical direction. At a frequency, ν, under the assumption of LTE, integrate over the region of atmosphere to find the total radiance at the top of the atmosphere:

$$L(\nu) = \int B(\nu, T)\, d\tau(\nu, P) \tag{1.14}$$

It is convenient to use as an altitude-dependent variable $z = -\ln P$, where P is the atmospheric pressure. In this case

$$L(\nu) = \int B(\nu, T)[d\tau(\nu, P)/dz]\, dz \tag{1.15}$$

In other words, the radiance L is the weighted average of the blackbody intensity; the weighting function is $d\tau/dz$, which is nearly independent of temperature. The weighting function represents the best region of the emitted radiation at a given frequency and height.

For a single frequency the altitude of the peak of the weighting function depends on the absorption coefficient at that frequency. For different frequencies, the weighting function will peak at different altitudes. Remote temperature sounding over a range of altitudes becomes possible if a set of frequency intervals can be chosen such that the corresponding radiances originate from substantially different levels in the atmosphere and the total height range represented includes all levels of interest.

Figure 1.4 shows the Nimbus–Satellite Infrared Spectrometer–B (Nimbus SIRS–B) weighting functions for different wavelengths in the 15-μm infrared band. Atmospheric linewidth, for instance in the 15-μm CO_2 band, varies from about 0.1 to 0.001 cm^{-1} over the range of atmospheric pressures involved. The bandwidth of a space sensor system would have to be better than 0.1 cm^{-1} in the wings of a line and as high as 0.001 cm^{-1} near the line centers. Many of the space radiometers designed for temperature sounding possess a spectral bandwidth that is much larger than this.

The most important gaseous constituents in the earth's atmosphere are water vapor, ozone, and carbon dioxide. Because of their strong absorption and emission bands, remote sounding to measure their distribution is possible. Water vapor and ozone have distributions that are very variable in time and place, and because of their significant interaction with the atmosphere, it is necessary to know their detailed distributions.

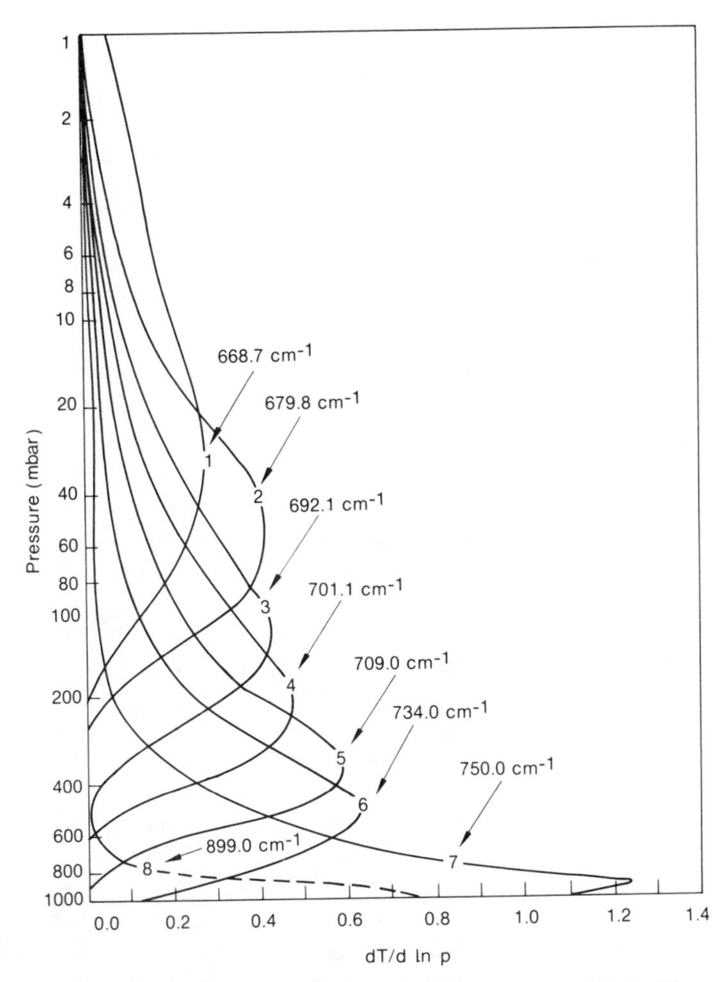

Fig. 1.4. Nimbus SIRS–B sensor weighting functions for each band in the 15-μm region. (Courtesy of D. Q. Wark, NASA.)

Remote sounding to determine the concentration of gaseous constituents may be carried out either through measurements of their emission bands in the infrared and microwave or through limb observations of their absorption of solar or atmospheric radiation transmitted or emitted by the atmosphere. For infrared emission measurements the temperature profile must be known, for example from measurements in the 15-μm band, where the effect of absorption by ozone and water vapor is relatively small. Then, in principle, additional radiance measurements in regions where the trace constituents are optically active will permit the recovery

Table 1.5

Space Sounding System Applications

Space sensor system	Applications
Passive UV	Ozone vertical profiles
Passive visible	Aerosol vertical profiles and upper atmospheric temperature and wind-field profiles
Passive IR and microwave	Vertical temperature, species concentration, and wind field
Active lidar	Vertical wind field, temperature, species concentration, and pressure
Active radar	Ocean surface wind field

of absorber amounts. Active lidar and microwave remote sounding have been considered for gaseous concentration measurements from space. Some success has been achieved with new technologies using these techniques, which are now far enough advanced to be considered for use on a satellite or a space station. Table 1.5 shows space sounding system applications.

A different approach to vertical space remote sensing of temperature and gas concentration profiles in the stratosphere involves the use of a space sensor looking at the earth's limb rather than straight down. Altitude sounding resolution is obtained by scanning the limb. A sensor employing a wide spectral band can be used to collect more radiation and partially offset the effect of the very narrow field of view.

The weighting functions are very sharp for limb-scanning sensors and always peak at the highest pressure in the field of view. Good vertical resolution (1 km) is obtained, while the horizontal resolution is near 10 km. Because of the increased opacity of the atmosphere when a tangent path near the limb is viewed, it is possible to sound considerably greater altitudes than with vertical sounding. Table 1.6 shows the altitude capability for each type of sounding.

For some limb sounding, solar occultation by the earth's atmosphere is

Table 1.6

Altitude Capabilities of Space Remote Sounding Methods

Space sounding method	Altitude (km)
Vertical sounding	0–100
Limb sounding	15–350

used. The space sensor tracks the bright solar disk, which can be used as a radiation source during each spacecraft sunrise and sunset event in order to detect an absorption profile through the earth's atmosphere.

Solar UV radiation backscattered from the ozone layer can be used to measure the vertical profiles and total amount of ozone in the earth's atmosphere. Grating spectrometers using multispectral techniques to measure backscattered solar UV radiation have been launched into space for remote sounding applications.

Figure 1.5 illustrates the space sounding techniques employed for (a) vertical atmospheric sounding, (b) atmospheric limb sounding, (c) solar occultation sunrise/sunset sounding, and (d) airglow and aurora emission sounding.

In the 1990s and beyond, many space sounding sensor systems will be developed in both the active and passive modes for applications related to the earth's climate, environment, and weather.

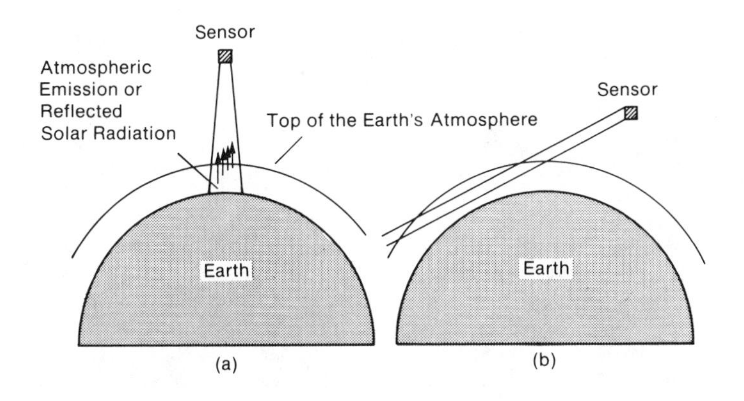

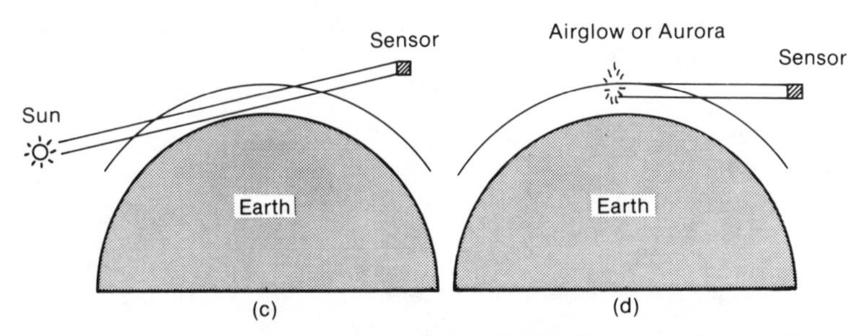

Fig. 1.5. Passive space sounding techniques. (a) Passive atmospheric nadir sounding technique, (b) passive atmospheric limb sounding technique, (c) solar occultation sounding technique, and (d) atmospheric emission sources sounding technique.

General References and Bibliography

Chamberlain, J. W. (1961). "Physics of the Aurora and Airglow." Academic Press, New York.

Clough, S. F., Knelizys, F., Gallery, W., Rothman, L., Abreu, L., and Chetwynd, J. (1982). FASCOD. *U.S. Air Force Geophys. Lab.* AFGL-TR-78-0081.

Coulson, K. L. (1975). "Solar and Terrestrial Radiation." Academic Press, New York.

Falcone, V. J., Abreu, L. W., and Shette, E. P. (1979). Atmospheric attenuation of millimeter and submillimeter waves: models and computer code. *U.S. Air Force Geophys. Lab.* **AFGRL-TR-79-0253.**

Fleagle, R. G., and Businger, J. A. (1980). "An Introduction to Atmospheric Physics." Academic Press, New York.

Fraser, R. S., and Curren, R. J. (1976). Effect of the atmosphere on remote sensing. *In* "Remote Sensing of Environment" (Joseph Lintz, Jr., and David S. Simonett, eds.), pp. 34–84. Addison-Wesley, Reading, Massachusetts.

Fritz, S., Wark, D. Q., Fleming, H. E., Smith, W. L., Jacobowitz, H., Hilleary, D. T., and Alishouse, L. C. (1972). Temperature sounding from satellites. *NOAA Tech. Rep., NESS* No. 59.

Gille, J. C., and House, F. B. (1971). On the inversion of limb radiance measurements. 1: Temperature and thickness. *J. Atmos. Sci.* **28,** 1427–1442.

Goody, R. M. (1964). "Atmospheric Radiation." Oxford Univ. Press, London and New York.

Houghton, J. T. (1977). "The Physics of the Atmosphere." Cambridge Univ. Press, London and New York.

Iqbal, M. (1983). "An Introduction to Solar Radiation." Academic Press, New York.

Kaplan, L. D. (1959). Inference of atmospheric structure from remote radiation measurement. *J. Opt. Soc. Am.* **49,** 1004–1007.

Kneizys, F., Shuttle, E., Gallery, W., Chetwynd, J., Abreu, L., Selby, J., Clough, S., and Fenn, R. (1983). Atmospheric transmittance/radiance computer code Lowtran 6. *U.S. Air Force Geophys. Lab.* **AFGL-83-0187.**

Liou, K. N. (1980). "An Introduction to Atmospheric Radiation." Academic Press, New York.

McClatchey, R., Fenn, R., Selby, I., Volz, F., and Garing, J. (1972). Optical properties of the atmosphere. *U.S. Air Force Cambridge Res. Lab.* **AFCRL-70-0527.**

Mateer, C. L., Heath, D. F., and Krueger, A. J. (1971). Estimation of total ozone from satellite measurements of backscattered earth radiance. *J. Atmos. Sci.* **28,** 1307–1311.

Smith, W. L. (1970). Iterative solution of the radiative transfer equation for the temperature and absorbing gas profile of an atmosphere. *Appl. Opt.* **16,** 306–318.

Thekaekara, M. (1976). Solar radiation measurement techniques and instrumentation. *Sol. Phys.* **18,** 309–325.

Twomey, S. (1963). Table of the Planck function for terrestrial temperatures. *Infrared Phys.* **3,** 9–26.

Vonder Haar, T. H., and Suomi, V. E. (1971). Measurement of the earth radiation budget from satellites during a five-year period. *J. Atmos. Sci.* **28,** 305–314.

Chapter 2 Spaceborne Radiation Collector Systems

2.1 Spacecraft Scanning Systems

Satellite spin scans from synchronous orbit for imaging and sounding applications have been done for many years. By rotating the spacecraft, the sensor field of view can be extended to a large area. The spin-scan axis can be perpendicular to the earth's surface or parallel to the earth's equator, as shown in Fig. 2.1.

A geosynchronous satellite could be placed in an inclined orbit. In this case, it would execute an elongated figure eight once a day, relative to the earth, centered on the equator and a fixed longitude, with the top and bottom of the figure eight at a latitude equal to the inclination of the orbit. A pushbroom space sensor can be used for this special orbit. Multi-figure eight inclined synchronous satellites, at about the same longitudes but with opposite phasing, would complement each other's coverage.

A further refinement would be an inclined eccentric orbit with a mean orbit altitude of 35,800 km. Because orbit velocity is inversely proportional to orbit altitude, the spacecraft would spend most of its orbit period near apogee (the highest point of the orbit) and relatively little near perigee (the lowest point), where it is moving fastest. An inclined eccentric synchronous orbit with both its apogee and time phase in a particular area would provide maximum continuous coverage of that area.

Pushbroom scanning is the technique of using the forward motion of a satellite platform to sweep a linear array of detectors oriented perpendicular to the ground track across a scene being imaged. The satellite motion provides one direction of scan and electronic sampling of the detectors in the cross-track dimension provides the orthogonal scan component to

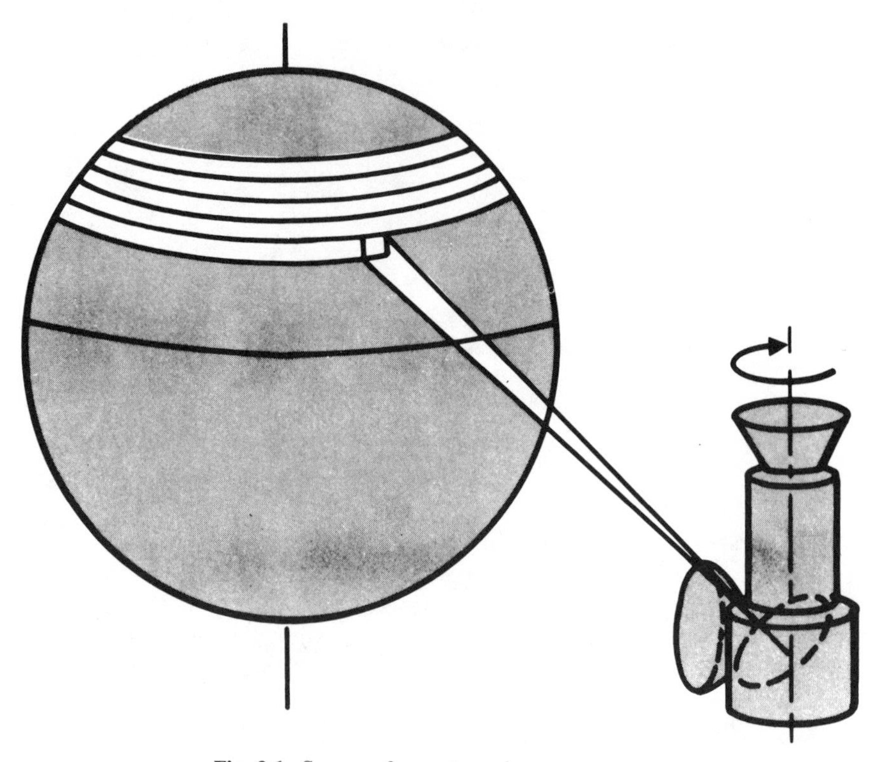

Fig. 2.1. Spacecraft equator spin-scan coverage.

form an image. The sensor detector array is sampled at the appropriate rate so that contiguous lines are produced. Figure 2.2 shows the satellite pushbroom scanning technique.

There are three principal advantages of pushbroom scanning techniques using long linear arrays of charge-coupled device (CCD) detectors. First, complex mechanical scan mechanisms are eliminated. Second, the longer dwell time offers a better signal-to-noise ratio. Third, the time delay and integration (TDI) technique can be applied.

Whiskbroom scanning is a technique employing a combination of the satellite's orbital motion and mechanical scanning transverse to the orbital motion. Figure 2.3 shows a whiskbroom sensor system imaging the earth.

2.2 Sensor Scanning and Pointing Systems

Space sensor imaging of a scene of interest is accomplished by scanning the field of view for large area coverage. Both mechanical and elec-

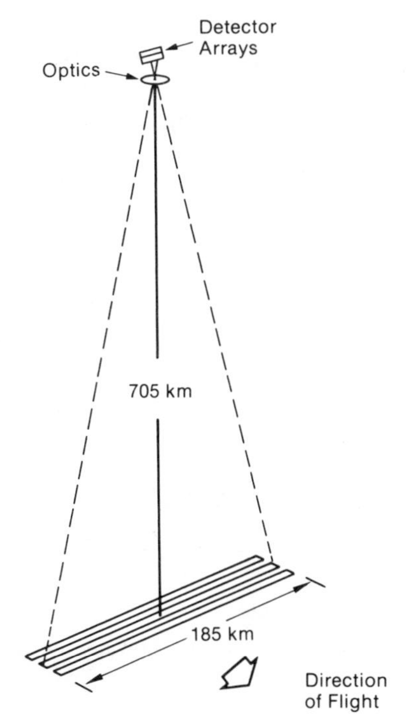

Fig. 2.2. Satellite pushbroom scanning technique.

tronic scanning techniques are used. In mechanical (whiskbroom) scanning, the field of view is changed by mechanical rotation of the sensor system or angular movement of the scan mirror either in the cross-track direction for the earth's surface or in a special direction for limb measurements. Electronic (pushbroom) scanning relies on the orbital motion of the sensor to sweep the area to be imaged.

The pushbroom sensor consists of an optical system having a wide field of view and thousands of detectors for each spectral band. No scanning mirror is required because the entire swath width is covered by the detector elements. Integrated focal plane arrays have been developed rapidly. Hybrid focal plane arrays have been assembled from CCD chips to cover a large field of view. Hence the pushbroom sensor is one of the best candidates for future space system applications.

Since the development of CCD detector arrays, large staring mosaic array sensors have been envisioned to replace some scanning systems. The performance advantages of staring mosaic systems include greater sensitivity because of their longer integration times. However, the uniformity, power, and sensor aperture sizing technology required for mosaic array sensors places them many years ahead of the available technol-

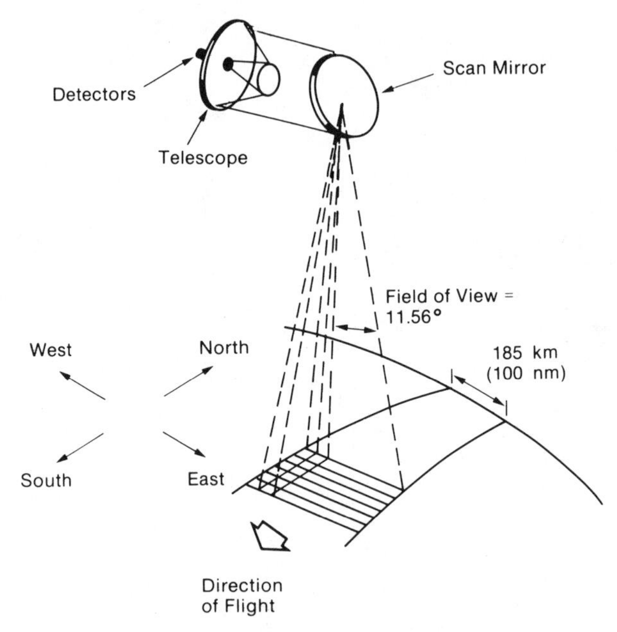

Fig. 2.3. Satellite whiskbroom scanning technique.

ogy. As the technology develops, mosaic array space sensors could replace some scanning sensor applications. Efforts are under way to reduce variations across arrays, to develop simple signal processing for uniformity correction, and to develop techniques for background signal suppression. If these efforts are successful, then staring mosaic operation of IRCCDs will be practical. Until then, most space applications are expected to use scanning devices for large field angle coverage.

Area arrays in a sensor scanning system include a parallel scanned array of detector channels to cover the field of view in the cross-scan direction and a serial array in each channel that employs time delay and integration (TDI) to achieve an improved signal-to-noise ratio in the scan direction. With TDI, the signals from individual detector elements in the scan direction are delayed and added in phase to achieve signal enhancement by increasing the number of elements in the serially scanned line. There are three advantages of TDI: (a) the signal-to-noise ratio is increased approximately by the square root of the number of elements; (b) the method provides an averaging effect that smooths out element-to-element nonuniformities and provides a better averaged uniformity than is afforded by a straight parallel scan array; and (c) a dead pixel or weak element in the string is held up by the others in the string.

With a TDI CCD detector, the space sensor will be capable of collecting a more useful signal. An additional consideration is that the introduction of a variable number of TDI integrations offers an alternative to the use of a step neutral density filter wheel to control the light intensity. The use of a single-element detector in a scanner in geosynchronous earth orbit (GEO) or low earth orbit (LEO) is very wasteful of the earth's radiation energy. With the TDI technique, better imaging and sounding results will be obtained with the advanced space sensor systems of the 1990s and beyond.

Some space sensor scanning systems require a three-axis gimbal system mounting for sensor pointing and tracking. Such a gimbal system, attached to a satellite, the shuttle, or a space station, will provide high-stability pointing of space sensor systems mounted upon it. A pointing system with stability in the range of 0.1 to 1.0 arc second has been required by space sensor users. The annular suspension pointing system (ASPS), which is being developed under NASA direction to satisfy user needs, is a modular system with two principal parts: an ASPS gimbal system (AGS), which has three conventional ball bearing gimbals, and an ASPS vernier system (AVS), which magnetically isolates the space sensor. Sensors requiring coarse to moderate pointing stability (30 arc seconds) can utilize the AGS, and those requiring the most stringent stability (0.1 arc second) can add on the AVS.

Another sensor pointing system is the instrument pointing system (IPS) of Spacelab. The IPS is a multipurpose precision pointing system (1.0 arc second) for sensors of various masses and dimensions.

2.3 Telescope and Receiver Systems

Many telescopes for space sensors make use of mirrors to form the primary image. A mirror has the advantage of being absolutely free from color aberrations. A spherical mirror has the big disadvantage of having spherical aberration. To overcome this, the spherical mirror is replaced by a parabolic or hyperbolic one. In a parabolic mirror all the light coming in parallel to the axis is focused in one point, and for this point the imagery is thus perfect. The field of view, however, is very small because coma is worse for a parabolic than for a spherical mirror. The usable field is of the order of not more than 1°.

Many space sensor telescopes make use of mirrors. The most common are the Cassegrain, Gregorian, Newtonian, Ritchey–Chretien, Schmidt, and Schwarzschild types. The Cassegrain type has a concave parabolic primary mirror, which collects parallel rays of light from the earth or other targets and concentrates them into a cone reflected in front of the primary mirror. Before the light focuses to a point, it is intercepted by a convex hyperbolic secondary mirror of smaller diameter, which spreads the cone and reflects it back through a hole in the center of the primary mirror. The desired images are obtained at the focus of the secondary mirror, which is the focal plane of the telescope. The primary advantage of the Cassegrain type is that a long focal length can be obtained in a short physical length, which means that less weight is required for space applications. The Gregorian design requires a longer tube, but can be designed for high-power application. Its concave secondary mirror is easier to use than the convex Cassegrain secondary. The Newtonian telescope has a parabolic primary mirror and a 45° flat mirror for small field angle applications. The Ritchey–Chretien configuration differs from the other types in that its primary and secondary mirrors are both hyperboloids. The use of hyperboloids for both mirrors allows simultaneous correction of third-order spherical aberration and third-order coma. The image quality of the Ritchey–Chretien is better than that of the classical Cassegrain type.

Schmidt and Schwarzschild telescopes are wide-field types for LEO pushbroom sensor and GEO space sensor systems. Progress in wide-field space sensor telescope system design over the past decade has been driven by parallel developments in large focal plane technology. The progress of CCD focal plane technology has demanded changes in sensor system requirements, including an increase in aperture size as smaller resolution has become of interest. Total fields of view are pressing toward 10 to 20° to accommodate earth coverage requirements. Focal planes are generally constrained to be flat field systems for ease of manufacture and hence lower cost.

A large number of design parameters are involved in a complete description of telescope design. These may be reduced to a small number of fundamental design parameters from which all the others may be derived. The choice of which parameters to consider fundamental is a matter of preference, since all are mathematically interrelated. In general, the three fundamental parameters of telescope design from a space sensor system point of view are focal length f, focal ratio (f-number) F, and entrance aperture diameter D. Since these three parameters are related by the equation

$$f = FD \tag{2.1}$$

only two of them need be considered fundamental. These quantities are shown graphically in Fig. 2.4.

Telescope diffraction occurs at the edges of the optical elements and at the aperture used to limit the incident beam. The image of a point source

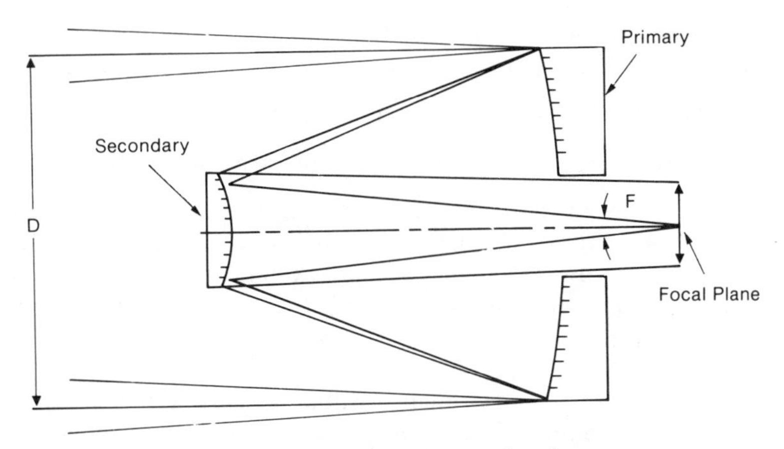

Fig. 2.4. Fundamental parameters of a telescope.

formed by diffraction-limited optics appears as a bright central disk surrounded by several alternately bright and dark rings. The central disk contains 84% of the radiance, with the rest in the surrounding rings. Since G. B. Airy was one of the first to analyze diffraction optics, the central disk is usually called the Airy disk. The angular diameter of this disk, which is considered to be equal to the diameter of the first dark ring, is given by

$$d = 0.244\lambda/D \tag{2.2}$$

where d is expressed in milliradians, λ in micrometers, and the aperture diameter D in centimeters.

In space sensor system design, the detector pixel size should be just large enough to extend to the center of the first dark ring. The pixel will then receive 84% of the radiance from the image.

The image of two targets consists of two diffraction patterns. If the targets are very close together, the images may overlap and become indistinguishable. Rayleigh suggested that the two images could be resolved if the center of one Airy disk coincided with the first dark ring of the other. This is called the Rayleigh criterion and is a good means of estimating the resolution capability of any diffraction-limited sensor system. Thus the minimum angular separation at which two point sources can just be resolved is

$$\alpha = 0.122\lambda/D \tag{2.3}$$

Equation (2.3) shows that for any diffraction-limited sensor system, whether it is an optical or a microwave sensor system, the angular resolution is proportional to the wavelength of the radiation and inversely proportional to the diameter of the aperture. For example,

$$\alpha = 10.0 \times 10^{-6} \quad \text{m}/1.0 \quad \text{m} \qquad \text{(infrared system)} \tag{2.4}$$

$$= 0.23 \quad \text{m}/2.3 \times 10^{4} \quad \text{m} \qquad \text{(microwave system)} \tag{2.5}$$

That is, for 10.0-μm radiation with a 1-m aperture system the angular resolution is equal to that for 0.23-m radiation with a 23-km large antenna system in the microwave region. Thus a synthetic aperture is needed in the microwave region to match the high resolution offered by sensor systems at optical wavelengths.

To maintain the surface of very large telescope mirrors, which are typically more than 3 m in diameter, within reasonable limits, a radical departure from traditional telescope manufacturing technology is required. Active telescope approaches, which allow the mirror figure to change in space, fall into two principal categories, corresponding to segmented and thin deformable mirrors. In the former, the mirror is cut into separate segments, each of which is moved as a rigid body until the optimum arrangement of the whole is reached. In the latter category, a single mirror is used, but it is provided with a large number of "push–pull" actuators in the rear to deform the surface into a diffraction-limited figure after orbital deployment. In each active approach, the actuator represents a complete and independent closed servo loop. In order to generate figure commands, surface sensing and computer systems are needed.

Table 2.1 shows the classification of telescope mirror systems. Figure 2.5 shows different telescope mirror systems used for space remote sensing applications.

Table 2.1

Telescope Classification

Telescope	Primary mirror	Secondary mirror
Cassegrain	Parabola	Hyperbola
Gregorian	Parabola	Ellipse
Ritchey–Chretien	Hyperbola	Hyperbola
Dall–Kirkham	Ellipse	Sphere
Newtonian	Parabola	Flat
Schmidt	Aspheric	Sphere
Schwarzschild	Sphere	Sphere (concentric)

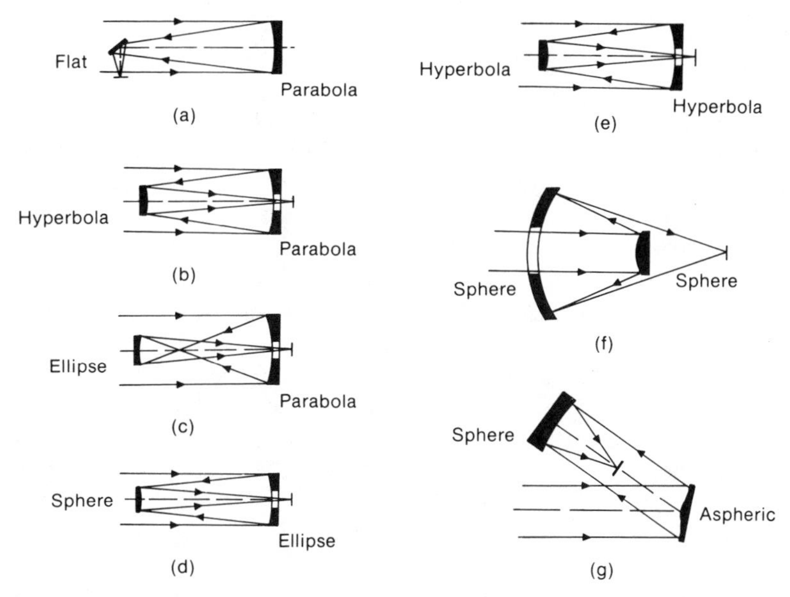

Fig. 2.5. Different space telescope systems: (a) Newtonian, (b) Cassegrain, (c) Gregorian, (d) Dall–Kirkham, (e) Ritchey–Chretian, (f) Schwarzschild, (g) Schmidt.

2.4 Space Sensor Mirror Materials

Because there is a need for better ground resolution for space sensor operation at longer wavelengths, large, lightweight mirror elements are required to maintain a near-diffraction-limited system. Materials used for lightweight mirror structures for space sensors will be discussed in this section.

Fused silica glass is a synthetic amorphous silicon dioxide of extremely high purity. A unique characteristic is its resistance to internal breakdown when exposed to gamma rays or electron bombardment. Other important physical characteristics are excellent elasticity, low thermal conductivity, and low dielectric losses. The average coefficient of thermal expansion per degree centigrade of fused silica is 0.56×10^{-6} over the range 0 to 300°C. This low thermal expansion value results in very high resistance to thermal shock, permitting fused silica to withstand extreme temperature gradients. Fused silica mirror blanks can fused into a strong monolithic structure.

Ultra-low-expansion (ULE) titanium silicate, Corning code 7971, is a synthetic amorphous silica glass manufactured by Corning Glass Works.

Developed primarily as an optical mirror material, ULE glass has the homogeneity, visible clarity, and low density of fused silica, but exhibits a coefficient of thermal expansion that is essentially zero over a wide temperature range. In addition, ULE glass is highly transparent, rigid, and polishable, making it the logical material for a first surface mirror. Most important, ULE glass can be formed into very large blanks of lightweight design, using the fusion techniques developed by Corning. The lightweight mirror blank configurations developed by Corning afford weight reductions ranging from 50 to 75% over comparable solid blanks. These reductions have proved valuable in space sensor systems, where weight saving is important. The original design employed a core of slotted struts, fitted together, with solid surface plates sealed to the struts. A new Corning technique allows plates of ULE glass to be fused together to form a monolithic lightweight core. Surface plates are then fused to the built-up core, and the resultant monolithic structure offers about 10% more rigidity than the equivalent slotted-strut design. The surface of lightweight blanks can be plano–plano, plano–concave, or convex–concave. Diameters over several meters are available.

Beryllium metal has afforded space sensor designers an ideal material that permits high-stiffness and low-weight structural designs. However, beryllium did not become a serious design candidate until the high cost of the raw material could be balanced against the stringent weight economics demanded by space applications. Because of its high stiffness-to-density ratio, one of the first space applications of beryllium was in high-speed rotating space sensors requiring relatively low moments of inertia. On the other hand, basic considerations such as material availability, cost, manufacturing, and metallurgical stability must not be neglected. The toxicity of beryllium in the form of salt, vapors, small particles, or dust is another drawback.

A beryllium powder metallurgy technique that results in random grain orientation has been developed and proved by use in many mirrors, including large ones. This process currently gives an upper limit of 1.6 m to the diameter of the primary mirror.

The usual approach with metal mirrors has been to deposit an electroless nickel coating on the figured surface, and to form the figure and polish on the nickel. For bare polishable grades of beryllium, the standard procedure has been to apply a hard, 0.075–0.175-mm-thick coating of electroless nickel to the heat-treated beryllium blank and prepare a high-quality optical surface on the nickel coating, which is significantly easier to polish.

Zerodur is the name given to a glass–ceramic material developed by Schott. Because of its almost negligible thermal expansion, high Young's modulus, and excellent polishing qualities, its main application is in the field of mirror optics.

Cervit is an Owens-Illinois material, essentially a glass which is changed to a polycrystalline solid by a proprietary process of nucleation and controlled thermal treatment. The C-101 composition developed for high-quality mirror optics has a very low thermal coefficient. Polishability equal to that of fused silica is claimed due to its small crystal size.

Table 2.2 shows a comparison of physical properties at 300 K for the different mirror materials.

2.5 Space Sensor Mirror Coating

Most space sensor mirrors have been coated by the technique of evaporating metal thin films. The spectral reflectance of most metal thin films increases at longer wavelengths. For the great majority of applications the coating material is an aluminum thin film, deposited on a mirror substrate by evaporation in vacuum. Aluminum has a broad spectral band of quite high reflectance and is reasonably durable when properly applied. If a sensor mirror is exposed to a harsh environment, it is necessary to protect its surface with a thin evaporated coating of silicon monoxide or magnesium fluoride. This combination produces a first surface mirror that is rugged enough to withstand regular handling and cleaning without scratching or other signs of surface wear. Table 2.3 shows the spectral reflectance for various evaporated metal thin films.

Table 2.2

Mirror Materials

Material	Density (g/cm^3)	Modulus of elasticity ($N/cm^2 \times 10^6$)	Thermal conductivity (cal/cm sec °C)	Specific heat (cal/g °C)	Coefficient of thermal expansion ($1/°C \times 10^{-6}$)
Fused silica	2.20	7.0	0.0033	0.188	0.55
ULE	2.21	6.74	0.0031	0.183	0.03
Cer-Vit	2.50	9.23	0.004	0.217	0.1
Zerodur	2.52	9.20	0.0039	0.196	0.05
Beryllium	1.86	28.0	0.38	0.45	12.4
Aluminum	2.70	6.9	0.53	0.215	23.9
Invar	8.0	14.8	0.026	0.095	1.3
Graphite epoxy	1.72	6.89	0.10	0.23	1.0

Table 2.3

Reflectance of Evaporated Metal Films

	Reflectance (%)			
Wavelength (μm)	Aluminum	Silver	Gold	Copper
0.22	91.5	28.0	27.5	40.0
0.24	91.9	29.5	31.6	39.0
0.26	92.2	29.2	35.6	35.5
0.28	92.3	25.2	37.8	35.5
0.30	92.3	17.6	37.7	33.6
0.32	92.4	8.9	37.1	36.3
0.34	92.5	72.9	36.1	38.5
0.36	92.5	88.2	36.3	41.5
0.38	92.5	92.8	37.8	44.5
0.40	92.4	94.8	38.7	47.5
0.45	92.2	96.6	38.7	55.2
0.50	91.8	97.7	47.7	60.0
1.0	94.0	98.9	98.2	85.0
5.0	98.4	98.9	98.6	98.7
10.0	98.7	98.9	98.9	98.9
15.0	98.9	99.0	99.0	99.0
20.0	99.0	99.2	99.0	99.2

2.6 Space Sensor Structural Materials

Space sensor systems involve structural elements that must have a high degree of rigidity, have good resistance to thermal deformation, and be light in weight. The structural materials most commonly used in current space sensor systems are aluminum and graphite/epoxy, with beryllium being used in specialized areas requiring isotropic high specific stiffness components. Within this group of materials, graphite/epoxy composites are perhaps the most versatile in their capability for being tailored for stress distribution, elastic modulus, and controlled coefficient of thermal expansion. However, shortcomings exist in the epoxy resin matrix potential for outgassing and moisture absorption, these shortcomings can contribute to space sensor contamination and possibly to dimensional changes. To meet future requirements for space sensor systems, metal-matrix composite materials are being considered. These materials are composites for high-modulus graphite fibers usually reinforcing either an aluminum or magnesium matrix. Attractive characteristics of metal-

Table 2.4

Comparison of Sensor Structural Materials

Characteristics	Rating of Materials				
	Al	Be	Gr/Ep	Gr/Al	Gr/Mg
Light weight	Fair	Fair	Good	Good	Good
High modulus	Poor	Good	Fair	Good	Good
No outgassing	Fair	Fair	Poor[a]	Fair	Fair
Electrically conductive	Fair	Fair	Poor	Fair	Fair
Cost	Low	Medium	Medium	High	High

[a] Fair with coating.

matrix composites are very good specific stiffness, excellent thermal conductivity, good electrical conductivity, near zero thermal expansion, no outgassing problem, and no moisture absorption. Table 2.4 shows a comparison of state-of-the-art space sensor structural materials.

The newer composites that are candidate structural materials for space sensors are graphite/aluminum and graphite/magnesium. In addition, carbon–carbon low-pressure processing modifications under development that will make low-cost structural shapes practical offer more possibilities for space sensor components, as do a series of relatively new composites based upon graphite fiber-reinforced glass matrices.

General References and Bibliography

Altenhof, R. R. (1976). Design and manufacture of large beryllium optics. *Opt. Eng.* **15,** 265–273.

Epstein, L. C. (1967). An all-reflection Schmidt telescope for space research. *Sky Telescope* Apr., 204–207.

Gratz, W. C. (1978). Progress in wide-field infrared systems design. *Proc. Soc. Photo-Opt. Instrum. Eng.* **156,** 83–86.

Hammesfahr, A. E. (1981). Instrument pointing subsystem (IPS) design and performance. *Proc. Soc. Photo-Opt. Instrum. Eng.* **265,** 117–125.

Loytty, E. Y. (1970). Mirror blanks for orbiting. *Opt. Spectra* Jan., 57–59.

Mika, A. M., and Richard, H. L. (1984). Optical-system design for next-generation pushbroom sensors. *Proc. Soc. Photo-Opt. Instrum. Eng.* **481,** 13–23.

Nummedal, K. (1980). Wide-field imagers—pushbroom or whiskbroom scanners. *Proc. Soc. Photo-Opt. Instrum. Eng.* **226,** 38–52.

Riper, R. V. (1981). High-stability Shuttle pointing system. *Proc. Soc. Photo-Opt. Instrum. Eng.* **265,** 134–153.

Simmons, G. A. (1970). Lightweight mirror blanks for space. *Opt. Spectra* June, 47–51.

Sletten, C. J., and Blacksmith, P. (1965). The paraboloid mirror. *Appl. Opt.* **4,** 1239–1251.
Smith, W. J. (1966). "Modern Optical Engineering." McGraw-Hill, New York.
Taylor, P. (1975). Metal mirrors in the large. *Opt. Eng.* **14,** 559–567.
Walter, G. D., ed. (1978). "Handbook of Optics." McGraw-Hill, New York.
Willey, R. R. (1962). Cassegrain-type telescope. *Sky Telescope* Apr., 191–193.
Yoder, P. R. (1953). Analysis of Cassegrainian-type telescopic systems. *J. Opt. Soc. Am.* **43,** 1200–1204.

Chapter 3 Space Detector Systems

3.1 Space Sensor Detectors

3.1.1 *UV and Visible Photomultiplier Tube Detector*

The photomultiplier tube (PMT) is a very useful space detector for UV and visible radiation detection both in passive and laser space sensor remote sensing application. The working principle of most PMTs is as follows: earth radiation falling on the photocathode causes it to emit free electrons, which are drawn away from the photocathode by dynodes having a more positive potential. These photoelectrons are then focused by various means on a secondary emission stage. Each primary electron striking this secondary stage frees more electrons. This process is then repeated, each stage having a more positive potential than the previous one. The electrons emitted from the last dynode are collected at an anode and the resulting amplified current is passed to the signal resistor and amplifier for recording and data processing systems.

There are two types of PMT: the end-window photocathode type and the side-window photocathode type. The most useful one for space applications is the end-window type. Figure 3.1 shows a schematic of an end-window PMT. It comes with various diameters from 1 cm up. The signal can be improved from 30 to 90% by introducing a prism on the window face so that for earth radiation from a particular direction there is total internal reflection in the glass face of the photocathode. Multireflection at the window admits more of the incident radiation to the photocathode for better signal collection. The Landsat multispectral scanner (MSS) sensor

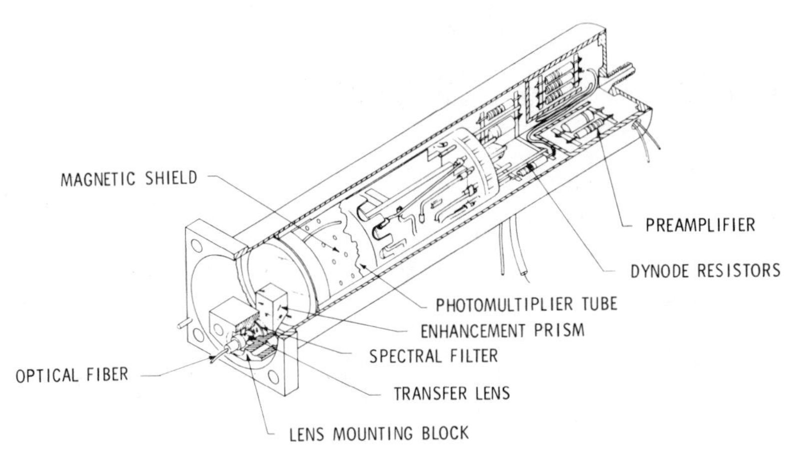

Fig. 3.1. Landsat photomultiplier tube system schematic. (Courtesy of Santa Barbara Research Center.)

uses the enhancement prism technique for better sensor design. Figure 3.2 shows the spectral sensitivity of the PMT.

The signal-to-noise ratio for a PMT is given by

$$S/N = i/(2qi\ \Delta f)^{1/2} \tag{3.1}$$

where i is the signal current, q the electronic charge, and Δf the bandwidth.

3.1.2 *UV and Visible Photodiode Detectors*

A silicon photodiode will generate a current proportional to the radiation incident on the active surface area. This photon-induced current, or photocurrent, will be divided between the diode internal junction resistor and the parallel load resistor. The photocurrent flow in the external resistor will produce a voltage across the load resistor. It is this signal voltage that will be amplified and recorded for data processing in the sensor system. A basic design question is whether to use a silicon photodiode in the unbiased photovoltaic (PV) mode or the back-biased photoconductive (PC) mode. Because the photovoltaic mode has no dark current, it has distinct advantages for low-level dc radiation signals. For ac chopped or pulsed light signals, the biased photoconductive mode is superior. The photovoltaic response time is limited to a few microseconds, while the back-biased photoconductive mode responds within a few nanoseconds.

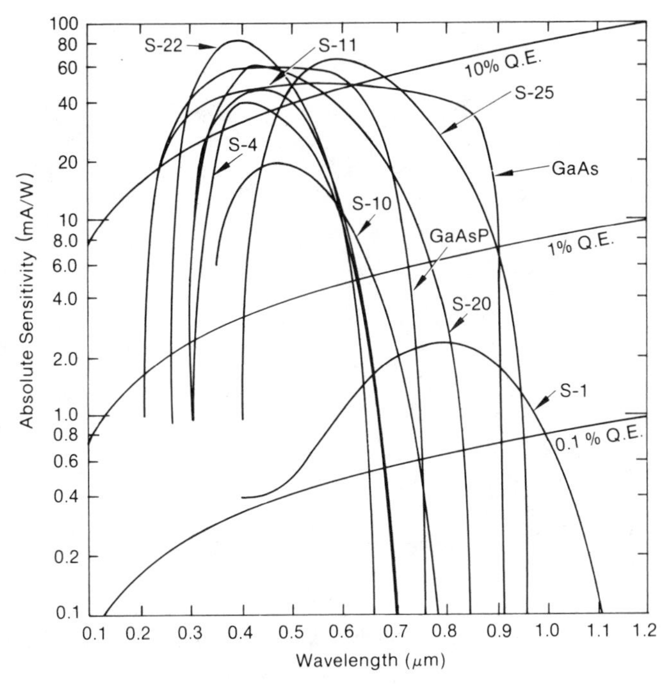

Fig. 3.2. Photomultiplier tube spectral responsivity. S-number is the code name of the spectral response. Q.E. stands for quantum efficiency.

Figure 3.3 shows the spectral sensitivity of silicon photodiodes. The signal-to-noise ratio of a silicon photodiode can be expressed as

$$S/N = i/\sqrt{\underbrace{(2qi\,\Delta f)^2}_{\text{shot noise}} + \underbrace{(4kT\,\Delta f/R)^2}_{\text{thermal noise}}} \tag{3.2}$$

where k is the Boltzmann constant, T the detector temperature, R the resistance of the load resistor, and i the signal current. Figure 3.4 shows the space photodiode system schematic of the Landsat MSS.

3.1.3 *Infrared Detectors*

The earth's infrared radiation can be measured either by observing the rise in temperature of a material used to collect the radiation or by measuring current effects due to the direct interaction of photons with electrons present in the detector. An infrared detector is characterized by its

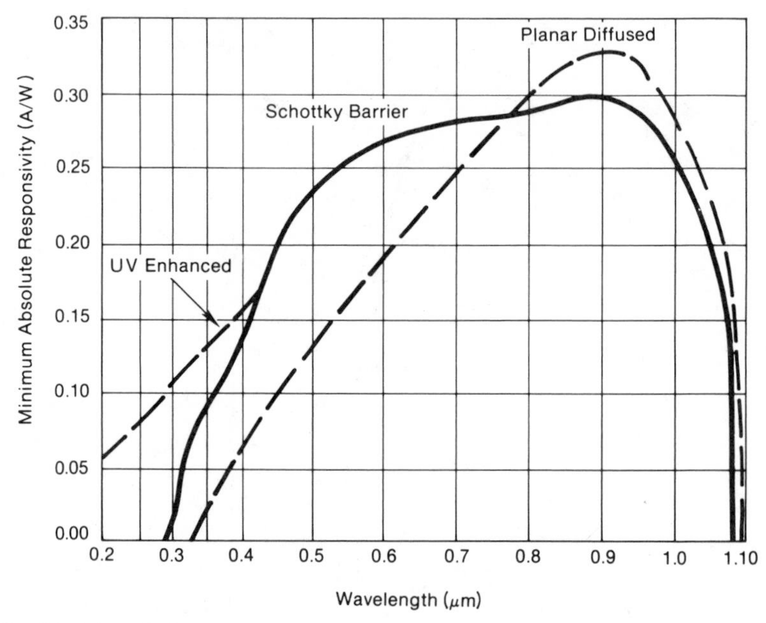

Fig. 3.3. Photodiode spectral responsivity. (Courtesy of United Detector Technology.)

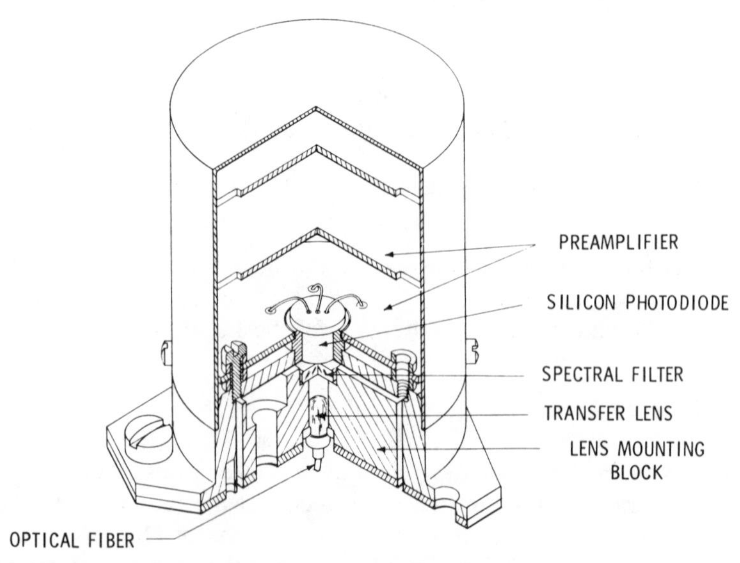

Fig. 3.4. Landsat MSS photodiode system schematic. (Courtesy of Santa Barbara Research Center.)

size, temperature, wavelength response, detectivity, frequency response, and background radiation.

The specific detectivity is a factor of merit D^*, which is the reciprocal of the noise equivalent power for a detector having unit area and a bandwidth of 1 Hz. It can be written as

$$D^* = \frac{(S/N)(A\ \Delta f)^{1/2}}{W} \tag{3.3}$$

where S/N is the signal-to-noise ratio, W the radiation incident on the detector, A the detector area, and Δf the bandwidth.

Table 3.1 lists the materials used for infrared detectors for space sensor systems, and Figure 3.5 shows the detectivity of infrared detectors as a function of wavelength.

Detector performance is usually characterized in terms of detectivity. An expression frequently used for the detectivity of near background-limited performance (BLIP) CCD detectors is

$$D^* = \eta\lambda/2hc(\eta E)^{1/2} \qquad \text{(photoconductive detector)} \tag{3.4}$$

$$D^* = \eta\lambda/\sqrt{2}\,hc(\eta E)^{1/2} \qquad \text{(photovoltaic detector)} \tag{3.5}$$

where D^* is the detector specific detectivity, λ the wavelength, h Planck's constant, c the speed of light, η the effective quantum efficiency, and E the number of background photons incident on the detector element.

3.1.4 *Microwave and Millimeter Wave Detectors*

Solid-state detectors and mixers used for the detection of microwave and millimeter wave radiation include point-contact Si or GaAs diodes and Schottky barrier Si or GaAs diodes. Point-contact diodes with a

Table 3.1

Intrinsic and Extrinsic Detector Materials

Intrinsic	Extrinsic
HgCdTe	Si : X
InSb	Ge : X
Si	GaAs : X
Ge	
PbS	
PbSe	
PbSnTe	

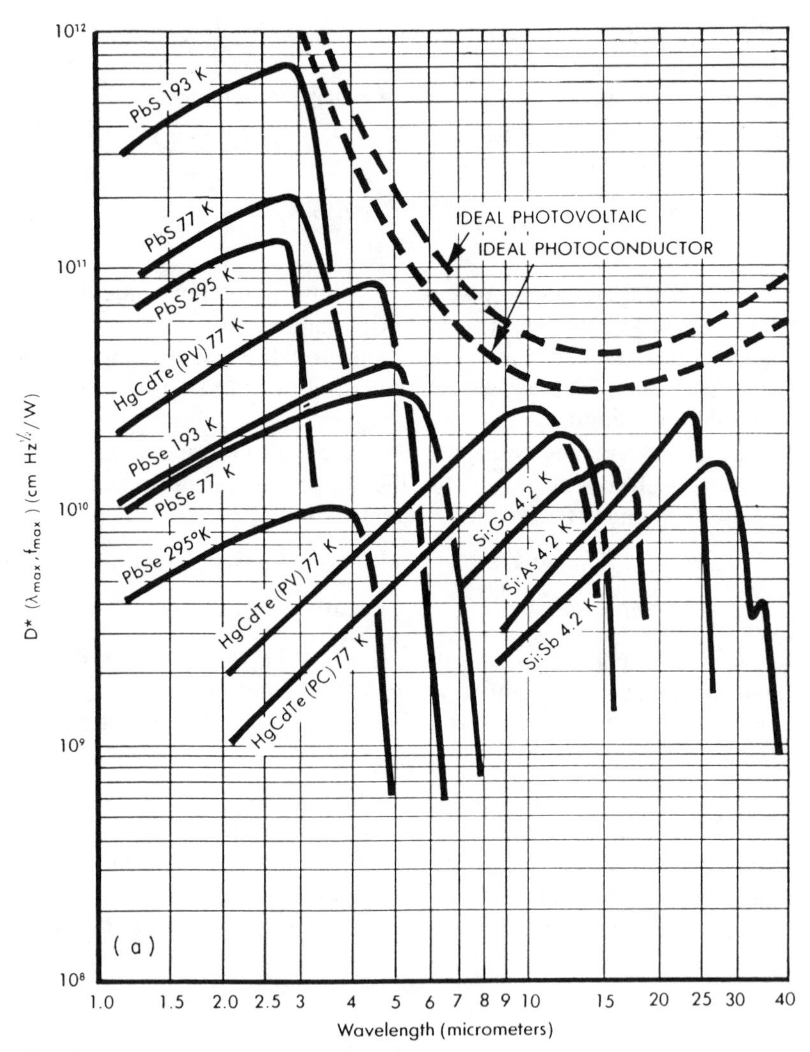

Fig. 3.5. (a) Infrared detector detectivities. (Courtesy of Santa Barbara Research Center.)

small-area pressure contact with the semiconductor surface have been used throughout the history of microwaves and millimeter waves. Point-contact diodes are fabricated with *n*-type semiconductor silicon or GaAs chips, contacted by a sharply pointed metal whisker, which is typically a chemically etched tungsten wire. Because of their low turn-on voltage and low cost, point-contact diodes are frequently used for microwave and millimeter wave detection. Schottky barrier diodes produced by planar

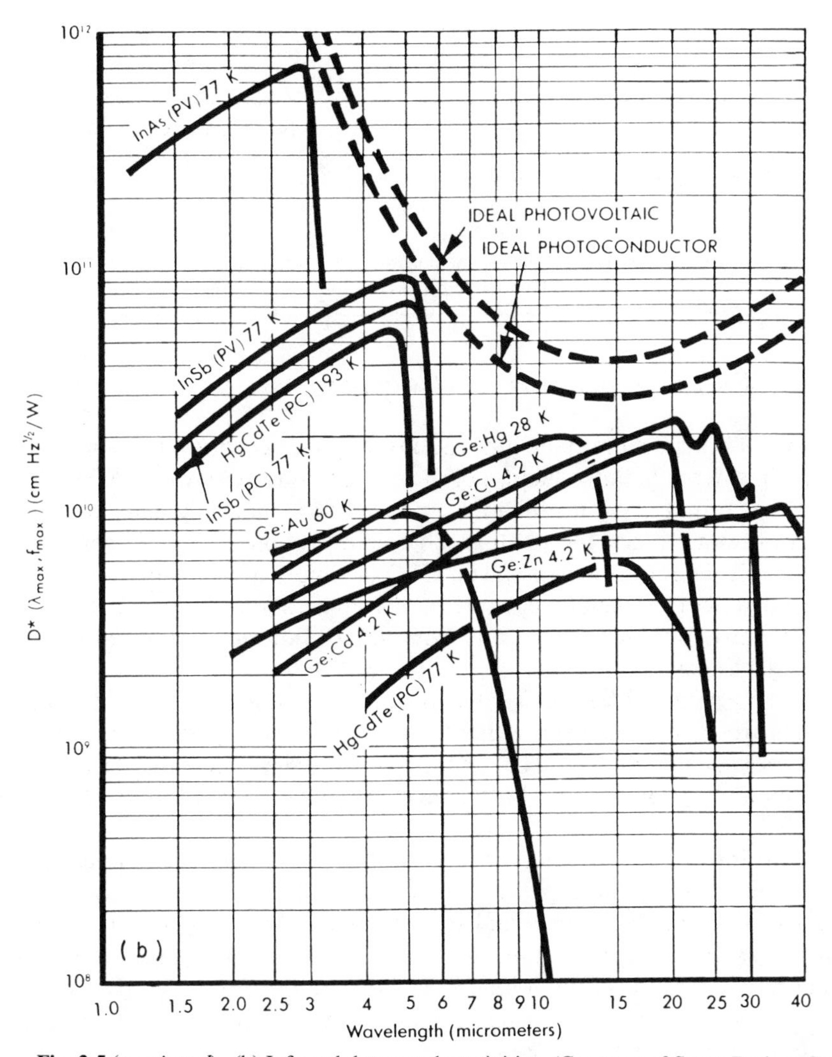

Fig. 3.5 (*continued*). (b) Infrared detector detectivities. (Courtesy of Santa Barbara Research Center.)

techniques have become more important for microwaves and millimeter waves. The Schottky diode has rectifying properties similar to those of a p–n junction, except that the former is a unipolar device in which current is carried only by the majority carriers. Schottky barrier diodes have a high degree of uniformity, obey Schottky theory almost perfectly, and have extremely low $1/f$ noise. Table 3.2 lists the cutoff frequencies of point-contact diodes and Schottky barrier diodes.

Table 3.2

Diode Cutoff Frequencies

Diode	Type	Cutoff frequency (GHz)
Point contact	*p*-type Si	76
	n-type Si	142
	n-type GaAs	1471
Schottky barrier	Beam-lead GaAs	796
	Honeycomb GaAs	1624

3.1.5 *Charge-Coupled Device Detector Systems*

One of the most dramatic innovations in space sensor systems in recent years has been the introduction of the charge-coupled device (CCD). The CCD has found many useful applications in space imaging and sounding, where it can replace the photographic emulsion plate, the vidicon tube, and single detector elements. The CCD is simply a means of controlling the movement of signal electrons by the application of electric fields. It shifts a group of signal electrons from input to output without distorting the signal itself.

The operation of the CCD detector requires that incident photons liberate charge-carrier electrons either from atoms of the crystal lattice (intrinsic type) such as Si, HgCdTe, or InSb or from impurities that have been intentionally added to the host lattice (extrinsic type) such as Si : X or Ge : X.

Most intrinsic CCD detectors are available in two modes of operation. When the photoconductive (PC) mode is used, a power source is connected to a load resistance and the detectors. Any variation in charge-carrier electron concentration produces a variation in voltage across the load resistor. In the photovoltaic (PV) mode, a p–n junction separates the charge carriers of opposite sign produced by the incident signal and no external power source is required.

Two basic CCD focal plane technologies, namely the monolithic and hybrid types, were developed in recent years for advanced sensor systems. In monolithic CCD arrays, photon detection and multiplexing are performed on the same chip. In the hybrid CCD technology, photon detection and multiplexing are performed by two separate chips, which are eventually interconnected by evaporated leads or indium bumps.

Monolithic CCD technology has the advantages of having the simplest possible structure that makes the best use of the most advanced large-scale integrated (LSI) circuit technology and having low production costs.

Hybrid CCD technology makes it possible to optimize independently the detector and the readout register, and to select the best available technology for each; the main problems are thermal mismatch of materials and higher detector impedance to match the CCD input.

Hybrid CCD technology often combines intrinsic or extrinsic detectors with a silicon CCD. Monolithic CCD technology employs materials such as Si, HgCdTe, and InSb. Hybrid arrays can be made in Z-package or planar architecture. The Z package has the advantage of providing more space in the Z direction to locate on-focal-plane processing functions.

Monolithic intrinsic space detectors can be made by two basic approaches: CCD technology and charge injection device (CID) technology. The CID approach can be viewed as an intermediate step to hybrid technology with one interconnection per cell between the detector chip and the multiplexer. Charge injection devices include an array of metal insulator semiconductor cells which are $X - Y$ addressed for readout. Hence, an $A \times B$ detector requires $A + B$ interconnections, located at the periphery of the device, to the separate silicon processing chip.

One of the best developed monolithic CCD technologies is based on extrinsic silicon, with certain dopants in the photosensitive layer providing a range of cutoff wavelengths in the IR region. The greatest advantage of the doped silicon detector is its good detectivity due to the cold background in space applications. The drawback of these CCDs is in the requirement for a very low operating temperature of the detector.

The Pd_2Si Schottky barrier CCD detector technology is relatively mature and potentially offers significant advantages in yield and productivity since standard silicon CCD processing is used. Another important feature is the detector uniformity, which reduces the processing procedures. The Pd_2Si detector is sensitive to radiation in the 1.2–4.9-μm region.

Table 3.3 shows a survey of space sensor CCD detector technology.

The functioning of space sensor CCD detectors depends critically on the

Table 3.3

CCD Detector Technology

Detector material	Wavelength (μm)	Temperature (K)	Array size	Pixel size (mil)
Si	0.4–1.1	200–300	1000 × 1000	0.2–0.5
HgCdTe	1.0–26.0	80–200	128 × 128	2–10
InAsSb	2–8	40–200	128 × 128	2–10
Si : In	2–8	20–40	64 × 64	2–10
Si : Ga	7–16	20–40	64 × 64	2–10
PbS	2.5–3.0	130	1000 × 2	2–15

Table 3.4

Space Detector Parameters, Symbols, and Units

Parameters	Units
Responsivity (*R*)	A W^{-1}
Detectivity (*D*)	W^{-1}
Specific detectivity (D^*)	cm $Hz^{1/2}$ W^{-1}
Quantum efficiency (η)	
Noise (*N*)	Electrons, V(rms), A(rms)
Signal-to-noise ratio (*S/N*)	
Noise-equivalent power (NEP)	W

focal plane temperature. As the spectral response extends to longer wavelengths, more cooling is required to reduce the number of carrier electrons liberated by incident signal photons and generated thermally.

Table 3.4 lists some useful detector parameters, including their symbols and units, for space sensor applications. Figure 3.6 illustrates the space remote sensing detector/CCD system schematic.

3.2 Visible and Near-IR CCDs

Silicon CCDs offer the capability for high-resolution, high-sensitivity, low-cost, high-reliability space sensors of minimal size, weight, and power consumption. They are expected to replace vidicon and other devices for space imaging applications.

Significant improvements are expected in intrinsic noise reduction and broader spectral bandwidth capability. Present silicon CCDs have a bandwidth covering 0.4 to 1.1 μm. As other materials are adopted, the limits will be extended below 0.2 μm and can reach about 18 μm for doped silicon devices.

A silicon CCD imager consists of an array of photosensitive elements, each of which is connected to a CCD element through an electronic gate. Light striking a photosensitive element produces free electrons by means of the photoelectric effect. After a suitable integration time has elapsed, the gate opens and the free electrons flow into the CCD. Then the gate closes and the accumulation of electrons begins again. Meanwhile, the CCD shifts the electrons one element at a time to an amplifier for signal output.

The technology for the preparation of monolithic focal plane arrays is fairly well established for visible and near-infrared detectors utilizing intrinsic silicon. Pushbroom imaging concepts are being used in the visible part of the spectrum, and are best illustrated by the NASA multispectral

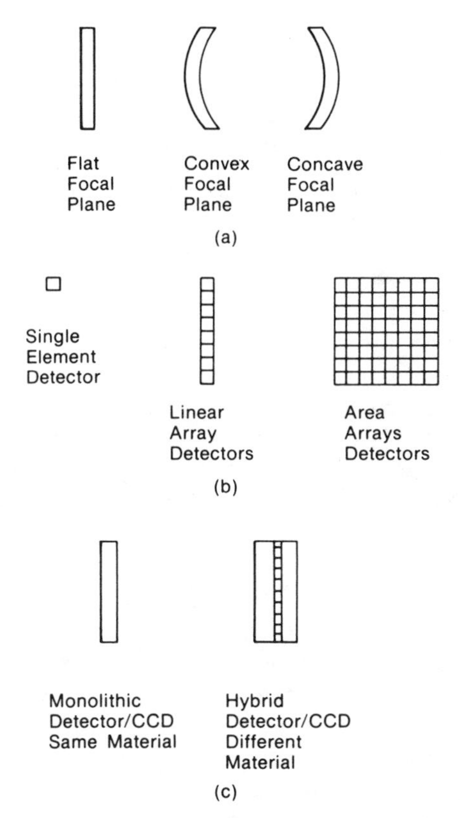

Fig. 3.6. Space detector/CCD system schematic: (a) side view, (b) front view, (c) side view.

linear array (MLA) program, the Satellites Probatories d'Observations de la Terre (SPOT)–high-resolution visible (HVR) CCD sensor, the German modular optoelectronic multispectral scanner (MOMS), and the Japanese marine observation satellite (MOS-1) sensor, all of which belong to the new generation of space sensor systems.

Critical parameters of silicon CCDs include noise and array size. Table 3.5 shows projected noise levels and CCD array sizes through the year 2000.

3.3 Infrared CCDs

For space applications involving infrared radiation, there are three spectral domains of interest, the currently accepted terminology being: short-wavelength infrared (SWIR), middle-wavelength infrared (MWIR),

Table 3.5

Noise Levels and Si Array Sizes

Year	Noise (e)	Linear array size	Area array size
1985	25–20	5,000	1,000 × 1,000
1990	20–15	7,000	2,000 × 2,000
1995	15–10	9,000	3,000 × 3,000
2000	5–2	10,000	4,000 × 4,000

and long-wavelength infrared (LWIR). The SWIR region covers reflected solar infrared radiation and is particularly useful in distinguishing earth surface features of interest. The MWIR region corresponds to the peak of thermal emission from targets and natural atmospheric emission. The LWIR region is centered on the peak thermal emission of the earth and is the region most commonly used for thermal imaging and vertical atmospheric temperature sounding.

Progress in the past few years on HgCdTe infrared CCD detector arrays hybridized with silicon CCD multiplexers has been very good. The feasibility of multiplexing large numbers of detectors on a focal plane has been clearly demonstrated, and large focal planes are likely in the future.

The use of HgCdTe detectors coupled to a silicon multiplexer appears especially attractive. For a given spectral region, HgCdTe CCD detectors afford the highest operating temperature. Table 3.6 shows HgCdTe detector operational parameters.

Room-temperature operation is reported for HgCdTe CCD detectors at wavelengths shorter than 2 μm. It appears that hybrid HgCdTe CCD arrays can be made to cover the SWIR, MWIR, and LWIR, three spectral ranges of interest for space applications. It is important that SWIR CCDs of this type are compatible with radiant or thermoelectric coolers, which makes them especially attractive for space sensor systems. For use in the MWIR, HgCdTe again stands out with regard to focal plane operating temperature, since 120 K is readily achievable with a passive cooler. No

Table 3.6

HgCdTe Detector Operational Parameters

Spectral range (μm)	Focal plane temperature (K)	Detectivity (cm $Hz^{1/2}$ W^{-1})
1–2.5	200	10^{12}
3–5	120	10^{11}
8–10	80	10^{10}

other detector technology seems to have more advantages than the HgCdTe technology.

Use of the LWIR spectral band for earth monitoring from space is well developed, and the availability of a good CCD system covering this spectral range is highly desirable. Unfortunately, it does not seem that any of the technologies yet developed is suitable, essentially because the cooling requirements of the system preclude the use of a passive cooler. Further efforts may improve this situation, and once again HgCdTe appears to offer the best solution.

Extension of HgCdTe to longer-wavelength response requires greater control of impurities for lower carrier concentrations in short-band-gap material. Infinite melt liquid-phase epitaxy seems to be the best solution for preparing better materials. There is a severe wavelength problem beyond 12 μm unless temperatures lower than 77 K are utilized and additional material and device development in this wavelength is pursued.

Of the several technical approaches to the development of high-density HgCdTe focal plane detector arrays for the IR region, the hybrid type represents a pragmatic approach that combines a mature detector technology with a well-developed silicon CCD multiplexer technology. Separation of the sensing and readout media allows the CCD benefits of high operating temperature, high quantum efficiency, and low crosstalk. The hybrid approach, however, does introduce a number of critical issues. First, an input circuit is needed to inject the detector signal into the CCD multiplexer, and this entails the expense of increased power dissipation. Second, a difference in thermal expansion coefficient between the detector material and silicon CCD can result in mechanical stress and poor reliability after temperature cycling; thus thermal expansion mismatch may place a limitation on the hybrid array chip size and require more chips per focal plane.

The performance of HgCdTe PC detectors at low frequencies is affected by $1/f$ noise, whereas that of PV detectors at high frequencies may be limited by amplifier noise. Photoconductive detector devices present a greater problem as more cooling is required due to the greater power dissipation in PC devices than in PV devices.

The targets of staring earth-view sensors produce fluxes at the detector that are much smaller than the background flux. The ratio of the background flux to the noise-equivalent input flux during an integration period is called the BNR and is the most important figure of merit for staring sensors. Because of photon statistics, high values of BNR require large photon fluxes on the detector. If each background photon produced an electron in the CCD multiplexer, the limited CCD well capacity, typically 10^6 charges, would severely limit the achievable BNR. Hence, many

Table 3.7

HgCdTe CCD Detector Array Parameters

Year	Number of detectors per linear array[a]	Linear array D^* (cm $Hz^{1/2}$ W^{-1})	Spectral range (μm)	Linear array detectors per centimeter
1980	20	10^{11}	1–16	100–200
1982	40			
1984	200			200–300
1986	500		1–22	
1988	1000			300–400
1990	2000	10^{12}		
1992	3000		1–26	400–500
1994	4000			
1996	5000			500–600
1998	6000			
2000	8000	5×10^{12}	1–30	600–800

[a] Stacked linear arrays.

methods have been demonstrated or proposed to prevent CCD saturation from limiting BNR performance. These methods include ac and dc suppression to reduce the background charge entering the CCD and rapid subframe averaging both before the CCD and in the signal processor.

Table 3.7 shows HgCdTe CCD detector array parameters projected to the year 2000.

3.4 Space Sensor Cooling Systems

In this section we shall consider the primary factors involved in the selection of an appropriate cooling system for new space applications. These factors are operating temperature, mission period, cooling capacity, weight, volume, and power availability. The various types of coolers can be classified in four categories: (a) passive radiant coolers, which cool systems to cryogenic temperatures by radiation to the low-temperature space environment; (b) open cycle, expendable systems, which use stored cryogens in either the subcritical or supercritical liquid state, solid cryogens, or stored high-pressure gas with a Joule–Thomson expansion; (c) closed-cycle, mechanical-refrigerator systems, which provide cooling at low temperatures and reject heat at high temperatures; (d) thermoelectric coolers, which use the Peltier cooling effect. Table 3.8 shows cooling systems for different temperature ranges for 3 to 5 years in a space sensor system.

Table 3.8

Cooling Systems for Different Temperatures

System temperature range (K)	Cooling
200–40	Passive radiant cooler or closed cycle
40–4	Closed cycle or open cycle

The systems most widely used so far for obtaining cryogenic temperatures aboard operational satellites have been passive radiant coolers. These systems are usually constructed in several stages. Passive space radiant coolers provide cooling by utilizing the low-temperature heat sink provided by the space environment. In a multistage radiant cooler system, each stage intercepts the heat load and radiates it into space from a surface whose area is inversely proportional to the fourth power of the absolute temperature. The characteristics of passive radiant coolers are: inherent reliability and simplicity, no mechanically moving parts, no vibration problem, no electrical power consumption, and proper spacecraft location and orbit orientation requirements.

Passive radiators have major applications in long-duration missions in either GEO or LEO for cooling to 70 K or above. At the lower temperatures small loads can be accommodated, while several watts can be dissipated at higher temperatures. The passive radiator concept is potentially attractive since the system requires no power and is capable of high reliability for extended periods. There has been considerable activity in the design of such radiators to maintain the temperature of CCD detectors in space sensor systems in the 70 to 150 K range. Currently, development efforts are under way to extent multistage passive radiant cooler performance down to 40 K.

For space sensor cooling in the 100 to 200 K range, passive radiant coolers provide the best approach for long life and design flexibility. The use of cryogenic heat pipe technology in passive radiator system design will broaden the cooling range of application. In the 40 to 100 K range, both mechanical refrigerators and radiant cooler multistage heat pipe systems are recommended for future applications. Below 40 K, mechanical refrigerators and open-cycle cryogenic systems (serviced by the Space Shuttle or a space station) have the potential for long-life applications.

The mechanical Vuilleumier (VM) refrigerator closed-cycle cooling system is a compact, high-performance cooler that can produce refrigeration at low temperatures for long periods of time without maintenance. It has a low wear rate on seals because the pressure difference across the dynamic seal in the machine is small. Thermal efficiency of the VM is relatively high because of its low mechanical friction. A practical VM refrigerator is capable of producing low temperatures to about 10 K. The VM refrigerator has three active volumes comprising a crankcase and two cylinders with displacers, which are arranged to cycle approximately 90° out of phase. A small timing motor drives the displacers in the correct phase state. Heat is supplied to the hot end of the power cylinder and rejected at the crankcase end of both cylinders. The expanded gas in the cold end of the refrigerator cools sufficiently to absorb heat from the cryogenic load. Developments in VM systems in the future should result in improved lifetime and efficiency.

Mechanical Stirling (MS) refrigerator closed-cycle cooling systems have had limited lifetimes for several reasons: (a) the limited life of dry lubricated bearings, (b) excessive vibration problems, and (c) seal wear causing working fluid leakage. The Stirling refrigerator contains no valves, compression and rejection of heat taking place in one cylinder with a heat exchanger, while heat absorption takes place in another cylinder. Development of a compact Stirling cooling system capable of operating for 3 to 5 years in space is needed. New Stirling closed-cycle systems have magnetic bearings and clearance seals, so that there are no rubbing surfaces, and have electronic control of axial position and piston and displacer phase angles. Contamination potential is minimized by eliminating all organics inside the working gas. A linear drive is used and the machine is dynamically balanced. It is expected that this new cooling system, with some improvements, could be made to last for as long as 5 years.

Other development programs for long-life mechanical coolers are being carried out by industry. These include rotary reciprocating coolers, turbo refrigerators, and various other machines. Each has attractive features, but none has demonstrated a sufficient lifetime. The development and verification of a long-life mechanical cooler is one of the most critical technological needs in the field of space cryogenic cooling.

In the absence of long-life mechanical refrigerators, the open-cycle or expandable cooling system approach, which is almost 100% reliable, is a viable alternative for low-temperature operation but is costly in terms of size and weight. Temperature ranges of open-cycle cooling systems employing the heat of sublimation or vaporization are 2 to 100 K. The limits of the cooling-temperature ranges for each coolant are based on a minimum defined by the solid phase at 0.1 torr pressure and a maximum defined approximately by the critical point. Cooling can be provided by selection of the proper coolant under the proper conditions. Table 3.9 shows operating temperature ranges for open-cycle coolants.

Table 3.9

Operating Temperature Ranges for Coolants

Cryogen	Operating temperature (K)
Helium	4.2
Hydrogen	20.3
Neon	27.1
Nitrogen	77.3
Argon	87.2
Oxygen	90.1

Open-cycle refrigeration systems include those using high-pressure gas with a Joule–Thomson (JT) expansion valve, cryogenic liquids in either the subcritical or supercritical state, and, for special applications, cryogenic solids. The advantages of these cooling systems are reliability, simplicity, relative economy, and negligible power requirements. In most cases, the technology is developed. Cryogenic fluids stored as liquids in equilibrium with their vapors can provide a convenient, constant-temperature control system. Fluids can be stored at pressures above their critical pressures (supercritical storage) as homogeneous fluids, thus eliminating the phase separation problems encountered during the weightless condition in space. The technology used in the liquid-storage systems has been extended so that lifetimes of approximately 3 years are feasible. A solid-cryogen system is used in conjunction with an insulated container, an evaporation path to space, and a conduction path from the coolant to the focal plane being cooled. Advantages over the use of cryogenic liquids include a higher heat content per mass and volume, lower operating temperatures, and no valves for pressure control. Solid cryogens provide a reliable refrigeration system for 3 to 5 years or longer in space cooling applications.

General References and Bibliography

Barbe, D. F., ed. (1980). "Charged-Coupled Devices." Springer-Verlag, Berlin and New York.

Blouke, M. M., Janesick, J. R., Hall, J. E., Cowens, M. W., and May, P. J. (1983). 800 × 800 charge-coupled device image sensor. *Opt. Eng.* **22,** 607–614.

Bode, D. E. (1980). Infrared detector. *Appl. Opt. Opt. Eng.* **6,** 323–355.

Cowley, A. M., and Sorensen, H. O. (1964). Quantitative comparison of solid state microwave detectors. *IEEE Trans. Microwave Theory Tech.* **MTT-14,** 588–602.

Danahy, E. L. (1970). The real world of silicon photodiodes. *Electro-opt. Syst. Des.* **5,** 36–43.

Hall, J. A. (1980). Arrays and charged-coupled devices. *Appl. Opt. Opt. Eng.* **8,** 349–400.

Hirschfeld, T. (1968). Improvements in photomultipliers with total internal reflection sensitivity enhancement. *Appl. Opt.* **7,** 443–449.

Hudson, R. D. (1969). "Infrared System Engineering." Wiley, New York.

Levinstein, H., and Mudar, J. (1975). Infrared detectors in remote sensing. *Proc. IEEE* **63,** 6–13.

RCA (1972). "Photomultipliers Manual." RCA, Harrison, N.J.

Sherman, A. (1978). Cryogenic cooling for spacecraft sensors, instruments, and experiments. *Astronaut. Aeronaut.* Nov., 39–47.

Santa Barbara Research Center (1982). "Infrared Components Brochure." SBRC, Goleta, California.

Tower, J. R., McCarthy, B. M., Pellon, L. E., and Strong, R. T. (1984). Visible and shortwave infrared focal plane for remote sensing instruments. *Proc. Soc. Photo-Opt. Instrum. Eng.* **481,** 24–33.

Wolfe, W. L., and Zissis, G. J., ed. (1978). "The Infrared Handbook." ERIM/Off. Nav. Res., Washington, D.C.

Chapter 4 Passive Space Radiometer Systems

4.1 Introduction

With the advent of the space age, our view of the earth's atmosphere and its surfaces was dramatically expanded. The space sensor view of the earth provided users with an exact image on a grand scale and with the capability for routine observation of the earth on a global basis. In April 1964, the first television and infrared observation satellite (TIROS) was launched for earth observation. The first Nimbus research and development satellite was launched in August 1964. The first application technology satellite (ATS) was launched in December 1966 to test advanced components and techniques for future geosynchronous satellite systems.

Many different space sensor systems have been designed for earth observation from space. So far, the radiometer has been the most useful sensor in space remote sensing applications. Space radiometers have certain sensor characteristics, such as high sensitivity, fast response, and relatively low design complexity and cost. A radiometer usually measures the radiance of electromagnetic radiation within a given spectral region. The simplest space radiometer system consists of a telescope, an optical interference filter, and detectors at focal plane locations. In a multispectral radiometer each radiometer filter has its own detectors at the same focal plane or different focal planes by use of a beamsplitter device. An alternative technique utilizes a single spectral radiometer and filters that are inserted sequentially into the sensor system. The sequencing can be accomplished with a rotating filter wheel containing fixed filters. The filter wheel technique reduces the number of detectors and sets of optical systems to one each, but at the expense of greater sampling time for the total system and an added mechanical subsystem.

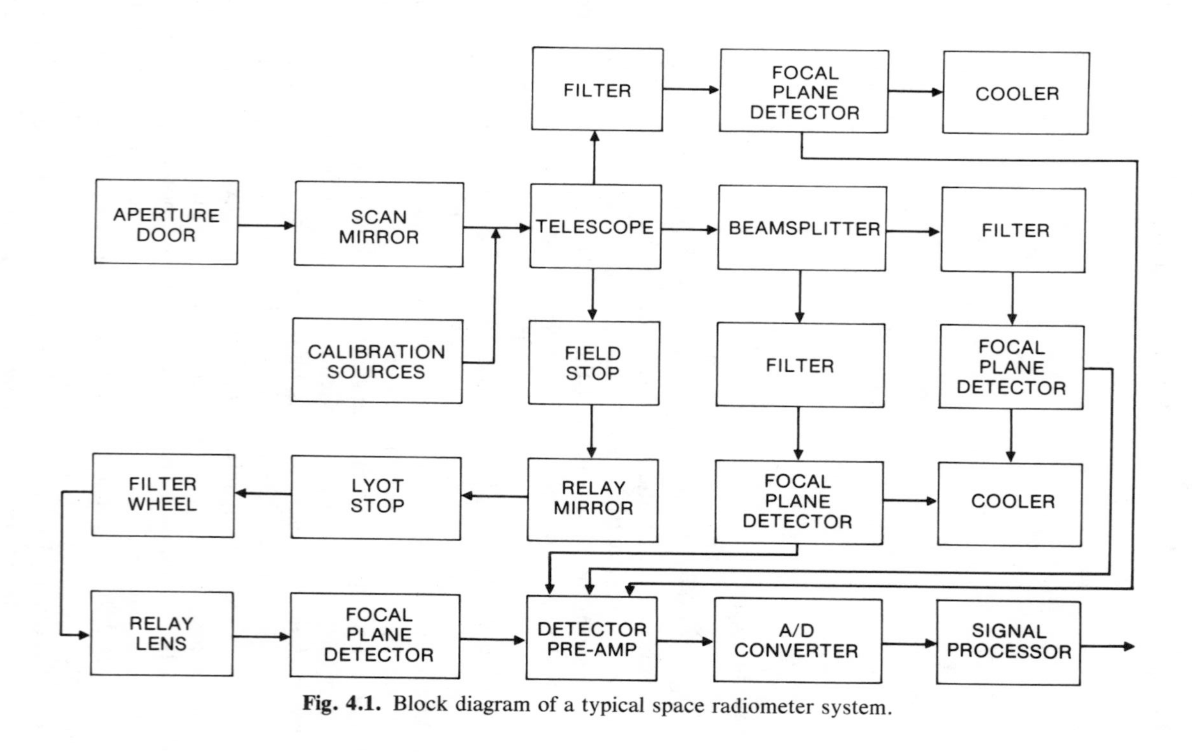

Fig. 4.1. Block diagram of a typical space radiometer system.

Table 4.1

Space Radiometric Quantities and Units

Quantities	Units
Radiant energy *(Q)*	J
Radiant density *(w)*	J/m^3
Radiant flux (Φ)	W
Irradiance *(E)*	W/m^2
Radiant intensity *(I)*	W/sr
Radiance *(L)*	$W/sr \cdot m^2$

Several types of measurements of the earth–atmosphere system are important in the UV, visible, and near IR; these are radiance measurements, polarization measurements, and time-dependent or viewing and solar angle measurements. Of importance in the infrared region are radiance measurements, cloud shadowing effect measurements, and radiation time-dependent measurements. Table 4.1 lists space radiometric quantities and units. Figure 4.1 shows a block diagram of a space radiometer system.

4.2 Radiometer Imaging and Sounding Spectral Band Selection

4.2.1 *Imaging Spectral Bands*

4.2.1.1 *Climate Sensor Spectral Band Selection*

The relationship between space observations and the earth's temperature, which is thought to be related to climate changes, is quite complex. A better understanding of this relationship will require more accurate radiation measurements such as those that can be obtained by the earth radiation budget (ERB) satellite system. The ERB satellite will provide measurements with sufficient spatial and temporal resolution to determine the earth's radiation budget on regional, zonal, and global scales over monthly, seasonal, and yearly time periods. The ERB sensor will permit the acquisition of more accurate radiation budget data than have been obtained in the past, and this will enable us to better understand climate and determine its predictability. The ERB sensor spectral bands are shown in Table 4.2. Spectral band selections for atmospheric gas measurements with future climate satellite sensors are shown in Table 4.3.

Table 4.2

ERB Spectral Band Selection

Channel	Spectral range (μm)	Applications
Starer		
1 Total	0.2–50+	Wide-FOV radiation
2 Shortwave	0.2–5	Wide-FOV radiation
3 Total	0.2–50+	Medium-FOV radiation
4 Shortwave	0.2–5	Medium-FOV radiation
5 Solar monitor	0.2–5	Solar radiation
Scanner		
6 Total	0.2–50+	Small-FOV radiation
7 Shortwave	0.2–5	Small-FOV radiation
8 Longwave	5–50+	Small-FOV radiation

4.2.1.2 *Land Observation Sensor Spectral Band Selection*

Remote sensing of the earth's surfaces from space has proved to be spectacularly successful in the Landsat space program. Many users throughout the world have found the data from Landsat's sensors to be a very useful source of information for applications and studies of land use. The spectral band selection of land observation sensors is shown in Table 4.4 and Table 4.5.

4.2.1.3 *Weather Space Sensor Spectral Band Selection*

Remote sensing from space of the earth–atmosphere system and its weather phenomena is one of the most mature space applications. The history of weather space sensors goes back to the beginning of the 1960s,

Table 4.3

Spectral Band Selections for Atmospheric Gas Measurements

Atmospheric gas	Spectral range (μm)	Applications
CO_2	4.5–5.33, 9.0–11.3	Greenhouse effect
NH_4	2.7–3.2, 8.0–14.0	Radiation effect
O_3	9.6–9.8	Radiation effect
N_2O	7.4–8.9	Radiation effect
CO	4.4–5.1	Radiation effect

Table 4.4

Landsat Thematic Mapper Spectral Band Selection

Spectral range (μm)	Applications
0.45–0.52	Costal area mapping; differentiation of soil and vegetation
0.52–0.60	Reflectance by healthy green vegetation
0.63–0.69	Chlorophyll absorption for plant species differentiation
0.76–0.90	Water body delineation; biomass surveys
1.55–1.75	Vegetation moisture measurements
2.08–2.35	Hydrothermal mapping; geology mapping
10.4–12.5	Thermal mapping; plant heat stress measurement

when TIROS sensors were launched and returned pictures of cloud cover to ground stations. Since those early days, there have been steady developments and improvements in space sensors. Most of these low earth orbit (LEO) weather space sensors were launched into sun-synchronous

Table 4.5

Land Observation Spectral Band Selection

Spectral range (μm)			
Landsat MSS[a,f]	SPAS MOMS[b]	SPOT HRV[c,e]	MOS MESSR[d]
0.5–0.6[a]	0.575–0.625[b]	0.50–0.59[c]	0.51–0.59[d]
0.6–0.7[a]	0.825–0.975[b]	0.61–0.69[c]	0.64–0.72[d]
0.7–0.8[a]	1.5–1.7[b]	0.79–0.90[c]	0.72–0.80[d]
0.8–1.1[a]	2.1–2.3[b]	0.50–0.90[e]	0.80–1.10[d]
10.4–12.6[f]			

[a] 76-m ground resolution.
[b] 20-m ground resolution.
[c] 20-m ground resolution.
[d] 50-m ground resolution.
[e] 10-m ground resolution.
[f] 234-m ground resolution.

polar orbit, so that each sensor passed over any location at the same local time. As the earth rotated beneath the sensor, total global coverage was achieved. Geostationary space sensors have also been utilized that are located in an orbit 35,800 km from the earth's surface. Rapidly varying weather phenomena can be observed every 30 min over the entire disk of the earth. Table 4.6 shows the spectral band selection and applications of environmental and weather space sensors.

4.2.2 *Sounding Spectral Band Selection*

Atmospheric sounders have been developed to obtain radiance data from which vertical temperature profiles of the atmosphere could be measured. April 14, 1969, is a historic date in the development of global vertical sounding techniques. On that date, Nimbus 3 carried two space sensors into orbit that provided space observations from which vertical temperature profiles of the atmosphere could be measured. The two space sensors are the satellite infrared spectrometer (SIRS) and the infrared interferometer spectrometer (IRIS). The satellite infrared spectrometer measured eight spectral intervals in the 11–15-μm spectral region, while the infrared interferometer sensor measured the infrared spectrum between 5 and 20 μm, from which water vapor and ozone, as well as the

Table 4.6

Environmental and Weather Sensor Spectral Band Selection

Spectral range (μm)	Applications
0.58–0.68	Cloud mapping
0.50–1.10	Surface albedo mapping
0.725–1.0	Surface boundaries
0.752–0.755	Clear atmosphere
0.759–0.762	Cloud height
0.761–0.763	Cloud height
1.059–1.219	Water vapor correction
1.548–1.70	Snow/cloud discriminator
2.05–2.28	Cloud particle size
3.55–3.93	Water vapor correction
5.70–7.00	Upper tropospheric wind field
9.60–9.80	Ozone total burden
10.30–11.3	Thermal mapping
11.50–12.5	Water vapor correction

Table 4.7

Sounder Spectral Band Selection

Spectral range (μm)	Applications
	Temperature
3.8–4.6	Vertical temperature profile
13.4–15.8	Vertical temperature profile
	Gas concentration
5.1–8.2	Vertical water vapor profile
0.25–0.34, 9.5–9.9	Vertical ozone profile

atmospheric temperature structure, have been obtained. Future space sounders will draw on new technology, and sounding systems in the 1990s will have improved resolution and will reach higher into the top of the atmosphere. The spectral bands of atmospheric sounders are shown in Table 4.7.

A big problem for high-resolution sounders is that interference from other gases may affect the radiance received by the sensor, and measuring vertical temperature profiles without monitoring the interference from atmospheric gases yields measurements of lower accuracy. A new space infrared spectrometer may be one of the best solutions for future advanced sounder systems.

4.3 Advanced Polarization Sensor Design and Analysis

Since the discovery by D. F. J. Arago in 1809 that the scattered light from the blue sky is polarized, the effects of various parameters on polarization have been observed and measured by many atmospheric scientists. All of these investigations have clearly indicated that aerosols and natural surfaces have large effects on this polarization, which therefore offers a powerful tool for monitoring the aerosol and natural surface properties in the earth–atmosphere system from space. Just like the space applications of the polarization of microwave radiation, space polarization measurement programs are needed in the visible and near-IR spectral regions for better monitoring of the earth's environment.

Radiance and polarization fields of the earth can be measured by a polarization space sensor system from a geosynchronous earth orbit (GEO) or LEO satellite system. Polarization of reflected and scattered solar radiation adds a new dimension to the understanding of the earth's environmental radiation field. It has been recognized that radiance and polarization space measurements are potentially useful, particularly for measurements of the condition of various earth surfaces, identification of cloud types, and determination of the turbidity of the atmosphere. With state-of-the-art technology, the radiance and polarization of the radiation emerging from the earth's surface and the atmosphere can be measured with a polarization space sensor system. The design and analysis of polarization sensors and measurements are described and discussed in the following sections.

4.3.1 *Theory of the Polarization Sensor*

In Fig. 4.2 the electric vector of electromagnetic wave propagation is in the direction $\hat{\mathbf{l}} \times \hat{\mathbf{r}}$. The vector can be written as

$$\hat{\mathbf{E}} = \hat{\mathbf{E}}_l + \hat{\mathbf{E}}_r = \mathbf{E}_l\hat{\mathbf{l}} + \mathbf{E}_r\hat{\mathbf{r}} \tag{4.1}$$

Let $\mathbf{l}$ be the component of the electromagnetic wave that is perpendicular to the scattering plane and $\mathbf{r}$ be the component that is parallel to the scattering plane; then $\mathbf{k}$ is the direction of propagation of the wave.

The electric vector components are

$$E_l = a_l \sin \psi = a_l i\, e^{-i\psi} \tag{4.2}$$

$$E_r = a_r \sin(\psi + \delta) = a_r i\, e^{-i(\psi+\delta)} \tag{4.3}$$

where

$$\psi = \omega t - \psi_l \tag{4.4}$$

$$\delta = \psi_l - \psi_r \tag{4.5}$$

The components of the electric vector after retardation has been introduced in the r axis are

$$E(\chi, \varepsilon, t) = E_l \cos \chi + E_r \sin \chi\, e^{i\varepsilon} \tag{4.6}$$

Letting angle brackets denote time averages, the intensity is

$$\begin{aligned} I(\chi, \varepsilon) &= \langle E(\chi, \varepsilon, t)E^*(\chi, \varepsilon, \tau)\rangle \\ &= \langle a_l^2 \cos^2 \chi + a_r^2 \sin^2 \chi + a_l a_r\, e^{i\delta} \sin \chi \cos \chi\, e^{-i\varepsilon} \\ &\quad + a_l a_r\, e^{-i\delta} \sin \chi \cos \chi\, e^{i\varepsilon}\rangle & (4.7) \\ &= \langle a_l^2(1 + \cos 2\chi)/2 + a_r^2(1 - \cos 2\chi)/2 \\ &\quad + a_l a_r \sin 2\chi(\cos \delta \cos \varepsilon + \sin \delta \cos \varepsilon)\rangle & (4.8) \\ &= \tfrac{1}{2}\langle(a_l^2 + a_r^2) + (a_l^2 + a_r^2) \cos 2\chi \\ &\quad + \sin 2\chi(2a_l a_r \cos \delta \cos \varepsilon + 2a_l a_r \sin \delta \sin \varepsilon)\rangle & (4.9) \end{aligned}$$

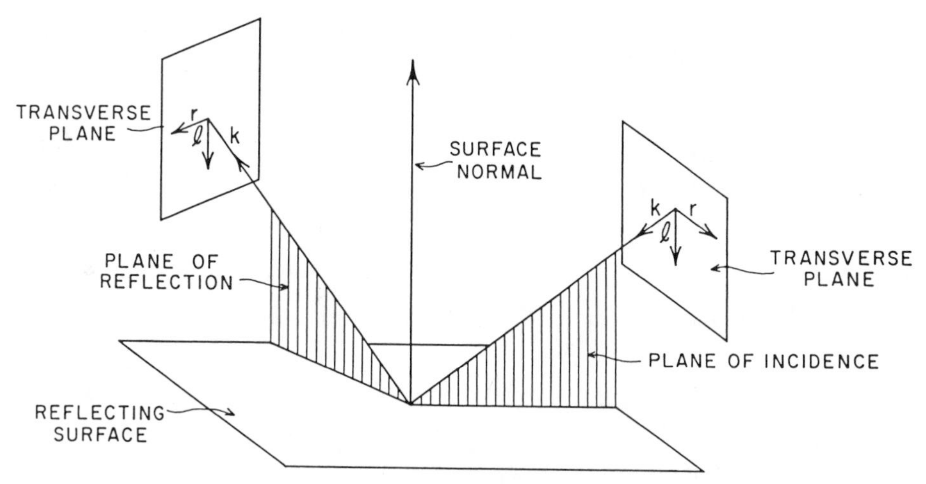

Fig. 4.2. Directional parameters for defining the Stokes vector.

Now I, Q, U, and V are defined as Stokes parameters by

$$I = \langle a_l^2 \rangle + \langle a_r^2 \rangle \tag{4.10}$$

$$Q = \langle a_l^2 \rangle - \langle a_r^2 \rangle \tag{4.11}$$

$$U = 2\langle a_l a_r \cos \delta \rangle \tag{4.12}$$

$$V = -2\langle a_l a_r \sin \delta \rangle \tag{4.13}$$

Then

$$I(\chi, \varepsilon) = \tfrac{1}{2}[I + Q \cos 2\chi + (U \cos \varepsilon - V \sin \varepsilon) \sin 2\chi] \tag{4.14}$$

One can determine the Stokes parameters by four set measurements such as $I(0°, 0°)$, $I(45°, 0°)$, $I(45°, 90°)$, and $I(90°, 0°)$. For the measurement of polarization of the scattered and reflected radiation emerging from the top of the atmosphere, if the measurement is restricted to linear polarization, three Stokes parameters (I, Q, and U) are required for the sensor system.

The polarization features of radiation characterized by the Stokes parameters can be determined by use of the following equations:

$$P = (Q^2 + U^2 + V^2)^{1/2}/I \tag{4.15}$$

$$\tan 2A = U/Q \tag{4.16}$$

where $\tan 2A$ gives the orientation of the plane of polarization, that is, the plane containing the major axis of the ellipse with respect to the l direction, and

$$\sin 2\beta = V/(Q^2 + U^2 + V^2)^{1/2} \tag{4.17}$$

where β, a measure of ellipticity, is given by arctan(b/a), and a and b are the semimajor and semiminor axes of the ellipse, respectively. In the case of linearly polarized light $V = 0$; thus the degree of polarization

$$P = (Q^2 + U^2)^{1/2}/I \tag{4.18}$$

When $\varepsilon = 0$, that is, in the absence of a retardation plate in the sensor system,

$$I(\chi, 0) = \tfrac{1}{2}(I + Q\cos 2\chi + U\sin 2\chi) \tag{4.19}$$

On the other hand, when $\varepsilon = \pi/2$, that is, when a quarter-wave plate is present in the sensor system,

$$I(\chi, \pi/2) = \tfrac{1}{2}(I + Q\cos 2\chi - V\sin 2\chi) \tag{4.20}$$

The polarization sensor system can be designed in accordance with the equations as follows for channels A, B, and C, respectively:

$$I(0°, 0°) = \tfrac{1}{2}(I + Q) \tag{4.21}$$

$$I(90°, 0°) = \tfrac{1}{2}(I - Q) \tag{4.22}$$

$$I(135°, 0°) = \tfrac{1}{2}(I - U) \tag{4.23}$$

It can be shown from the above cases that the Stokes parameters can be determined from the following equations:

$$I = I(0°, 0°) + I(90°, 0°) \tag{4.24}$$

$$Q = I(0°, 0°) - I(90°, 0°) \tag{4.25}$$

$$U = I(0°, 0°) + I(90°, 0°) - 2I(135°, 0°) \tag{4.26}$$

4.3.2 *System Design and Analysis*

4.3.2.1 *GEO System*

The polarization sensor is similar in operation to the GOES system. Satellite spin generates the west–east scan while precision stepping of the scanner generates the north–south scan. A major characteristic of this sensor is that the north–south scan is achieved by rotating the complete sensor around an axis at the spin axis. Each spin of the spacecraft at 100 rpm sweeps out scanning lines within the west–east frame limits of $\pm 10°$ from earth center. In the $20° \times 20°$ full earth frame there are 2550 lines for bands 1–6. The ground resolution at the satellite subpoint is about 5×5 km for bands 1–6. The 20° frame time is 25.6 min for each of the six bands. The data rate is 54.1 Mbps for each band.

Lightweight mirrors are required for the Ritchey–Chretien 24-in.-diameter telescope. Four arrays of six detectors per array for bands 1–3 at the primary focal plane and six detectors per array for bands 4–6 at the secondary focal plane form the optical field stop. Figure 4.3 shows the focal plane detector format.

Polarizers in different orientations are mounted in front of the detector arrays. One filter wheel is required. The filter wheel contains six spectral bandpass filters in the ranges 0.40–0.48, 0.48–0.58, 0.63–0.69, 1.17–1.30, 1.55–1.75, and 2.08–2.35 μm.

Table 4.8 gives the computed signal-to-noise ratio (S/N) results. The radiance for each band is calculated from the solar radiation, earth's atmospheric transmittance, and albedo of the surface. The detector responsivities presented in the table are the values found in publications of the detector suppliers. The total transmittance of the system is the product of the mirror reflectance, filter transmittance, and polarizer transmittance. The clear aperture is the net useful area of the primary, the secondary, and the mount's obscuration. The fixed noise current is made up of several noise sources. The noise values presented are based on typical measured values for the circuit design. The encoder and quantization $(S/N)_Q$ is calculated for the 10-bit system with $\frac{1}{2}$-bit encoder error. The value $(S/N)_T$ is the total signal-to-noise ratio of the sensor system.

4.3.2.2 *LEO System*

The pushbroom polarization sensor system is similar in orbital motion to the Landsat system, as shown in Figure 2.2. Spacecraft motion provides one direction of scan, and sampling of the pixels of the detector system in the cross-track dimension provides the orthogonal scan components to form an image. The detector array is sampled at the appropriate rate so that contiguous lines are produced. Three arrays are typically used for each spectral band for the radiance and polarization measurements. The system provides a 30 × 30-m detector footprint with 6300 detectors in each array. No scan mirror is required for this system. Total field of view is 15.38° across the track. Figure 4.4 shows the focal plane format of the sensor. The data rate is 51.5 Mbps per band.

Fig. 4.3. IFOV polarization focal plane format.

Table 4.8

Signal-to-Noise Ratio and Polarization Accuracy Computation

Band	Spectral bandwidth (μm)	Total low-level radiance (mW/cm^2 sr)	Detector	Detector responsivity (A/W)	Trans-mittance	IFOV (μrad)	Area (cm^2)	Noise bandwidth (kHz)	Noise current (pA)	*l*-direction polarization radiance (mW/cm^2 sr)	*r*-direction polarization radiance (mW/cm^2 sr)	$\left(\frac{S}{N}\right)_Q^a$	$\left(\frac{S}{N}\right)_T^b$	ΔP^c (%)
1	0.40–0.48	0.13	Si	0.20	0.18	139.6	2480	37.6	1.76	0.06		578	54	±0.6
			295K						1.84		0.07	675	62	±0.6
2	0.48–0.58	0.25	Si	0.31	0.24	139.6	2480	37.6	2.63	0.12		266	134	±0.3
			295K						2.71		0.13	288	143	±0.3
3	0.63–0.69	0.13	Si	0.44	0.24	139.6	2480	37.6	2.34	0.06		230	109	±0.3
			295K						2.46		0.07	269	122	±0.3
4	1.17–1.30	0.10	InSb	0.6	0.20	139.6	2480	37.6	4.19	0.06		400	77	±0.5
			95K						4.03		0.04	266	53	±0.5
5	1.55–1.75	0.08	InSb	0.8	0.20	139.6	2480	37.6	4.23	0.05		416	85	±0.5
			95K						4.03		0.03	250	53	±0.5
6	2.05–2.35	0.046	InSb	1.0	0.20	139.6	2480	37.6	4.03	0.024		279	53	±0.7
			95K						4.00		0.022	255	50	±0.7

[a] $(S/N)_Q$, quantization signal-to-noise ratio.
[b] $(S/N)_T$, total system signal-to-noise ratio.
[c] ΔP, polarization accuracy.

Fig. 4.4. Pushbroom scan.

A wide-field lightweight telescope is required for the sensor. At the focal plane there are three arrays of 6300 silicon detectors for bands 1–3, and three arrays of 6300 PbS or HgCdTe detectors for bands 4–6. Polarizers in three different orientations are mounted in front of the detector arrays. Two filter wheels are required. One filter wheel contains three spectral bandpass filters for filter selection in bands 1–3. The second contains three spectral bandpass filters for filter selection in bands 4–6.

Table 4.9 gives the computed signal-to-noise ratio and polarization accuracy analysis results. It is a system trade-off modeling computation. The quantization signal-to-noise computation is calculated for a 12-bit system with a $\frac{1}{2}$-bit error. Here $(S/N)_T$ is the total signal-to-noise ratio of the pushbroom polarization sensor system.

Geosynchronous earth orbit and low earth orbit observations from space have become a major factor in improving our understanding of the environment, new resources, and the climate of the earth. Development of the polarization sensor will consist of more improvements in existing space sensors. The polarization sensor system offers information on the following aspects of the earth and the environment: new resources location, soil moisture, crop production, pollution distribution, snow and ice surface conditions, and ice cloud and water cloud classification.

4.4 Gas Filter Analyzer System Design and Analysis

Nondispersive space sensors for remote sensing of temperature and atmospheric gas will be presented in this section. Nondispersive sensors use a reference cell containing an atmospheric gas as an optical filter, and the atmospheric radiation is modulated only at wavelengths where the atmospheric gas absorbs. The other cell is a vacuum cell. Atmospheric radiation passing through the vacuum cell is clearly not changed, whereas

Table 4.9

Signal-to-Noise Ratio and Polarization Accuracy Computation

Band	Spectral bandwidth (μm)	l-direction polarization radiance (mW/cm^2 sr)	r-direction polarization radiance (mW/cm^2 sr)	Detector	Responsivity R (A/W) or $D^* \times 10^{11}$ (cm Hz$^{1/2}$/W)	Trans-mittance	IFOV (μrad)	Area (cm^2)	Noise bandwidth (kHz)	$\left(\frac{S}{N}\right)_T$	ΔP (%)
1	0.40–0.48	0.06		Si	0.20R	0.36	42.5	729	0.114	90	$<\pm 1$
			0.07	295K						96	$<\pm 1$
2	0.48–0.58	0.12		Si	0.31R	0.48	42.5	729	0.114	100	$<\pm 1$
			0.13	295K						101	$<\pm 1$
3	0.63–0.69	0.06		Si	0.44R	0.48	42.5	729	0.114	92	$<\pm 1$
			0.07	295K						94	$<\pm 1$
4	1.17–1.30	0.06		PbS	0.61D^*	0.40	42.5	729	0.114	83	$<\pm 1$
			0.04	295K						56	$<\pm 1$
5	1.55–1.75	0.05		PbS	0.83D^*	0.40	42.5	729	0.114	90	$<\pm 1$
			0.03	295K						63	$<\pm 1$
6	2.05–2.35	0.024		PbS	1.0D^*	0.40	42.5	729	0.114	60	$<\pm 1$
			0.022	295K						56	$<\pm 1$

the radiation passing through the gas cell is altered. If the spectral signal of the radiation entering the cell does not match that of the gas, the difference in signal recorded by the detector between the radiation passing through the gas cell and that passing through the vacuum cell should be a constant. If the radiation signal shows absorption at the spectral lines of the gas cell of interest, then the signal difference should be a function of the emitted atmospheric radiation for temperature remote sounding or a function of the concentration of the gas for remote sensing of atmospheric gases. The gas filter analyzer has the advantages of high spectral resolution and large étendue or throughput.

4.4.1 *Theory of the Gas Filter Analyzer*

The detectivity of the gas filter analyzer system is inversely proportional to the noise-equivalent power and can be expressed as

$$D_s = D^*/\sqrt{A_d \, \Delta f} \tag{4.27}$$

where D^* is the specific detectivity, A_d the detector area, and Δf the noise-equivalent bandwidth.

The available radiant power at the detector focal plane can be written as

$$P \, \Delta\nu = A_0 \Omega L \, \Delta\nu \tau \tag{4.28}$$

where A_0 is the useful area of the entrance aperture of the sensor, Ω the solid angle of the instrument, $A_0\Omega$ the étendue or throughput, L the spectral radiance, $\Delta\nu$ the spectral resolution, and τ the system efficiency.

In order to establish the sensitivity of the sensor to changes in atmospheric radiance levels caused by temperature in the atmosphere, it is necessary to consider the total spectral radiance received by the space sensor. Thus,

$$\Delta P \, \Delta\nu = \tau A_0 \Omega \Delta L \, \Delta\nu \tag{4.29}$$

The signal-to-noise ratio is proportional to the system detectivity and the available radiant power integrated over the wavelength interval. For gas spectral lines with small wavelength intervals, S/N can be expressed as follows:

$$S/N = D_s P \, \Delta\nu \tag{4.30}$$

Combining all the equations, one can write

$$S/N = \tau A_0 \Omega D^* \, \mathrm{SC} \, L \, \Delta\nu/\sqrt{A_d \, \Delta f} \tag{4.31}$$

The dimensions of the various quantities are: D^*, centimeters hertz$^{1/2}$/watt; A_d, square centimeters; Δf, hertz; A_0, square centimeters; Ω, steradians; L, watts centimeter per square centimeter steradian; and $\Delta \nu$, reciprocal centimeters; τ, S/N, and SC are dimensionless.

Now let us evaluate each of the above quantities for the space gas filter analyzer system.

4.4.1.1 *Sensor Detector Specific Detectivity D^**

This quantity is defined as the reciprocal of the noise-equivalent radiance for a detector of 1-cm^2 area and a bandwidth of 1 Hz. Detectors having the highest D^* at the spectral interval of interest with high cooling temperatures were used for the space gas filter sensor design and analysis. The detectors used in the S/N calculation were HgCdTe for a range of 4.6 to 11.0 μm.

4.4.1.2 *Detector Area A_d*

Detector area is determined from telescope focal length, sensor altitude, and ground resolution. Typical values of the detector area range from 2 × 2 mil to 100 × 100 mil.

4.4.1.3 *Noise-Equivalent Bandwidth Δf*

The noise-equivalent bandwidth can be shown to be related to the dwell time of the instrument. For a gas filter space sensor, the usual relationship is

$$\Delta f = 1/2t_D \tag{4.32}$$

where t_D is the dwell time in seconds.

4.4.1.4 *System Efficiency τ*

System efficiency is a function of the number and quality of the electro-optical components (mirror reflectance, filter transmittance, etc.) and the atmospheric transmittance.

4.4.1.5 *Useful Area of the Sensor Aperture A_0*

The term A_0 is the area of the primary mirror minus the area of the secondary mirror or the net radiometric aperture area. The diameter of the primary will produce the diffraction-limited blur circle or the minimum image size. The larger the aperture the better the signal or the smaller the footprint that can be produced.

4.4.1.6 *Solid Angle of the Sensor* Ω

The solid angle is the angle subtended at the sensor by a ground area of specified size when the telescope is used. For square ground resolution $\Omega = 4 \sin^2 \alpha = (2\alpha)^2$ [where 2α is the instantaneous field of view (IFOV)], and for circular ground resolution $\Omega = \pi \sin^2 \alpha = \pi\alpha^2$.

4.4.1.7 *Spectral Resolution* $\Delta\nu$

The spectral resolution indicates the width of spectrum that the sensor can measure. For the gas filter analyzer, the band spectral resolution is determined by the useful portion of the absorption bandwidth of the gas in the cell and the filter used in the optical system. The absorption bandwidth of the gas and the number of useful lines in that bandwidth are obtained from an analysis of the Air Force Geophysics Laboratory (AFGL) atmospheric absorption lines tape. The useful portions are shown in Table 4.10.

4.4.1.8 *Spectral Radiance L*

The basic equation governing the magnitude of the upward infrared radiant energy received by the space sensor from a surface of temperature T and the polluted atmosphere can be written as

$$L(\nu) = \varepsilon B(T, \nu)\tau(T_s, P_s) + \int B_\nu(T, P) \frac{d\tau}{dz} dz \tag{4.33}$$

The Planck function for a blackbody B_ν is a function of temperature and wavelength. The transmittance of the polluted atmosphere is a function of

Table 4.10

Useful Portion of Spectral Bandwidth for Several Atmospheric Gases

Pollutant	Wavelength (μm)	Useful spectral bandwidth (cm^{-1})
CO	4.6	112
NO_2	7.6	96
CH_4	7.8	141
SO_2	8.7	132
NH_3	10.4	224

Table 4.11

Pollutant Concentrations in Rural and City Areas

Pollutant	Typical rural concentration (ppm)	Typical city concentration (ppm)
CO	0.1	5.0
NO_2	0.001	0.1
CH_4	1.4	3.0
SO_2	0.002	0.2
NH_3	0.01	0.01

the absorption coefficient, total path length, atmospheric pressure, pollution concentration profile, and temperature profile.

4.4.1.9 *Pollution Signal Change SC*

In order to establish the sensitivity of the instrument to small changes in radiance levels caused by a pollutant in the atmosphere, it is appropriate to consider the fractional change in radiance caused by the presence of the pollutant rather than the total spectral radiance received with the pollutant present. Thus, one can write

$$\Delta L = [(L_{\text{clean}} - L_{\text{poll}})/L_{\text{clean}}]L_{\text{clean}} = \text{SC } L_{\text{clean}} \tag{4.34}$$

Table 4.12

Gas Filter Analyzer Signal-to-Noise Ratios[a]

Pollutant	Wavelength (μm)	Infrared detector (77°K)	D^* (cm Hz$^{1/2}$/W)	Spectral radiance (W/cm^2 sr cm^{-1})	Signal change	Useful spectral width (cm^{-1})	TDI gain	$\frac{S}{N}$
CO	4.6	InSb	1.2×10^{10}	2.2×10^{-8}	2.0×10^{-2}	112	3	8
NO_2	7.6	HgCdTe	0.8×10^{10}	2.1×10^{-6}	2.1×10^{-4}	96	3	5
CH_4	7.8	HgCdTe	0.9×10^{10}	2.3×10^{-6}	4.6×10^{-2}	141	1	>100
SO_2	8.7	HgCdTe	1.2×10^{10}	2.8×10^{-6}	4.5×10^{-4}	132	3	30
NH_3	10.4	HgCdTe	1.6×10^{10}	6.4×10^{-6}	1.3×10^{-2}	224	1	>100

[a] Results are for the following conditions: satellite altitude, 900 km; ground resolution, 1 × 1 km; aperture size, 1 × 10^4 cm^2; solid angle, 1.2 × 10^{-6} sr; system efficiency, 0.1; dwell time, 0.1 sec; and detector area, 0.01 cm^2.

Table 4.11 lists typical pollutant concentrations in rural and city areas used for modeling considerations.

The signal-to-noise ratio results given in Table 4.12 are for pollution sounding from a satellite at 900-km LEO altitude.

4.5 Space Radiometer Sensors and Applications

4.5.1 *LEO Imaging and Sounding Applications*

4.5.1.1 *Climate, Environment, and Weather Applications*

The first series of LEO satellite systems designed for atmospheric studies was named TIROS in the early 1960s. The Nimbus class is a series of experimental environmental satellites.

Nimbus 1 carried the first high-spatial-resolution radiometer when it was launched in August 1964. The purpose of this sensor was infrared imaging, using an IFOV equivalent to a resolution of 8 km at the earth's surface. The spectral band was selected in the 3.4–4.1-μm atmospheric window. This window was chosen in preference to those at longer wavelengths because it was the only one for which better infrared detectors were available at that time. The earth pictures were built up by scanning the field of view from horizon to horizon using a rotating scan mirror turning uniformly at a rate of 447 rpm. This rotation rate was chosen so that the satellite would advance just enough between each swath to provide contiguous coverage of the ground. This sensor measured medium-infrared earth emission for day and night applications. Space measurements of the infrared maps of clouds contain information about cloud-top temperature and cloud height.

Nimbus 4 was launched in April 1970 with a temperature and humidity infrared radiometer as one of its payloads. The longer-wavelength spectral band at the 10–12-μm atmospheric window was selected so that measurements could be taken both day and night near the maximum region of the earth's infrared emission. A second spectral band at 6.5–7.0 μm was selected to observe atmospheric emission near the center of the water vapor absorption band.

An improved TIROS operational satellite system (ITOS), the second-generation operational environmental satellite system, was launched in October 1972 and carried three main sensors: a scanning radiometer (SR) and a very high resolution radiometer (VHRR), each with two spectral channels, one in the visible at 0.5–1.1 and 0.6–0.7 μm and one in the infrared 10.5–12.5-μm window with an HgCdTe detector cooled by a radiant cooler to near 105 K, and a vertical temperature profile radiometer (VTPR). The VTPR is a filter wheel radiometer with eight filters (535, 668.5, 677, 695, 708, 725, 747, and 835 cm^{-1}) covering atmospheric CO_2 15-μm band emission from the surface up to 20 km for vertical atmospheric temperature profile measurements.

The third-generation operational environmental satellite system, TIROS-N/NOAA, was placed in service in October 1978. The primary environmental space sensors for the satellite system are as follows.

The high-resolution infrared radiation sounder (HIRS) is a 20-spectral-band space sensor making measurements primarily in the infrared region of the earth's emitted radiation. The sensor is designed to provide data that will permit radiative transfer calculation of the vertical atmospheric temperature profile from the surface to 30-km altitude, vertical water vapor profile, and total ozone content. Table 4.13 shows the HIRS sensor specifications.

Table 4.13

HIRS Sensor Imaging and Sounding Applications[a]

Channel	Central wave number (cm^{-1})	Bandwidth at 50% transmittance (cm^{-1})	Wavelength (μm)	Applications
1	688.5	3.0	14.96	Vertical
2	680.0	10.0	14.71	temperature
3	690.0	12.0	14.49	sounding
4	703.0	16.0	14.22	
5	716.0	16.0	13.97	
6	733.0	16.0	13.64	
7	749.0	16.0	13.35	
8	900.0	35.0	11.11	Thermal
9	1030.0	25.0	9.71	mapping
10	1225.0	60.0	8.16	Vertical
11	1365.0	40.0	7.32	water vapor
12	1488.0	80.0	6.72	sounding
13	2190.0	23.0	4.56	Vertical
14	2210.0	23.0	4.52	temperature
15	2240.0	23.0	4.46	profile
16	2270.0	23.0	4.41	
17	2360.0	23.0	4.24	
18	2515.0	35.0	3.98	
19	2660.0	100.0	3.76	Thermal mapping
20	14500.0	1000.0	0.69	Imaging

[a] Courtesy of NOAA.

Table 4.14

SSU Sensor Sounding Applications[a]

Channel	Central wave number (cm^{-1})	Cell pressure (mbar)	Weighting function peak (km)
1	668	100	29
2	668	35	37
3	668	1.5	45

[a] Courtesy of NOAA.

The stratospheric sounding unit (SSU) is a gas filter-type sensor employing selective spectral absorption of CO_2 gas to make measurements in three gas filters. The spectral characteristics of each channel filter are determined by the pressure in a CO_2 gas cell in the sensor optical path. The condition of CO_2 in the cells determines the height of the weighting function peaks in the earth's atmosphere from 29- to 45-km altitude. Table 4.14 shows SSU sensor specifications.

The advanced very high resolution radiometer (AVHRR) is a five-band scanning radiometer in the visible, near-infrared, and infrared regions for earth surface, cloud, and thermal mapping imaging applications. Table 4.15 shows AVHHR sensor specifications.

Figure 4.5 shows the HIRS weighting functions and Fig. 4.6 illustrates the footprints as function of scan angle for the HIRS and scanning microwave spectrometer (SCAMS) sensors.

The importance of the earth's radiation budget in determining climate has long been recognized. The earth radiation budget space sensor provides the capability for simultaneously monitoring the outgoing solar radi-

Table 4.15

AVHRR Sensor Imaging Applications[a]

Channel	Spectral band (μm)	Applications
1	0.58–0.68	Cloud mapping
2	0.725–1.0	Surface water boundaries
3	10.3–11.3	Thermal mapping
4	3.55–3.93	Water vapor correction
5	11.5–12.5	Water vapor correction

[a] Courtesy of NOAA.

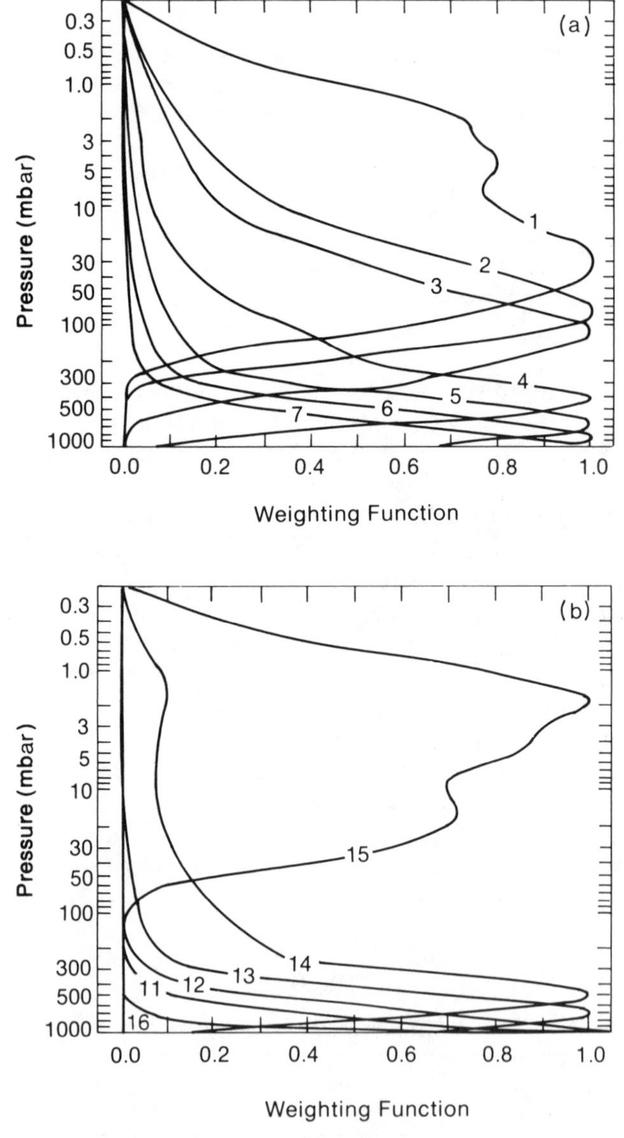

Fig. 4.5. Weighting functions for the HIRS temperature channels. (a) Longwave (channels 1–7), (b) shortwave (channels 11–16). (After Smith *et al.*, 1975.)

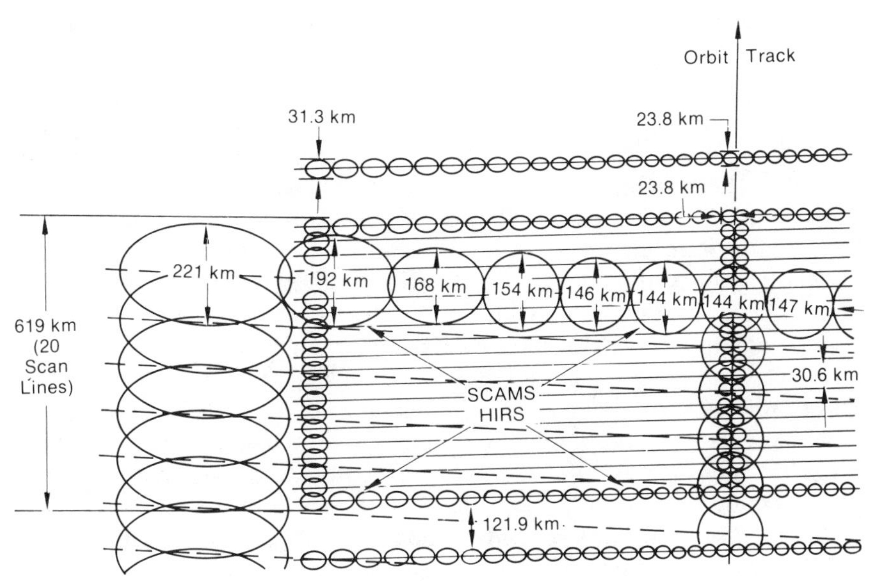

Fig. 4.6. Scan grid patterns for HIRS and SCAMS.

ation reflected from the earth's surface and atmosphere and the longwave radiation emitted by the earth and the atmosphere. The sensor system is capable of monitoring these fluxes on a daily, monthly, seasonal, or yearly basis for extended periods of time. These capabilities make it possible to conduct detailed studies of the variations of the earth's radiation budget, the effects on this budget of natural and man-made changes in the environment, and the effects on the earth's weather and climate produced by changes in the energy budget.

A basic component of the radiation budget of the earth is the input from solar radiation or the solar constant. Just how constant this is as a function of time and space has been a matter of study for a long time. Accurate measurements from the surface are difficult because of variations in atmospheric transmittance. Therefore, space is the best place to measure the solar constant.

For the average radiation balance at the top of the atmosphere between the solar radiation input and the outgoing reflected solar radiation and emitted radiation, one can write

$$I_0 \pi r^2 (1 - a) = 4\pi r^2 \varepsilon \sigma T^4 \tag{4.35}$$

where I_0 is the solar constant, a the earth's albedo, T the earth's mean surface temperature, ε the earth–atmosphere system emissivity, σ the

Stefan–Boltzmann constant, and r the earth's radius. If we let $a = 0.33$, $\varepsilon = 0.66$, and $I_0 = 1370\ \mathrm{W/m^2}$, then $T = 280$ K.

The earth radiation budget satellite system (ERBSS) will be used to study problems of great importance to climate research. In order to determine the cause of radiation changes, careful measurements of cloud size and cloud height must be associated with earth radiation budget measurements. A microwave earth radiation budget sensor would be added to the system for better data correlation analysis.

The gas filter radiometer or selective chopper radiometer is used to determine atmospheric vertical temperature profiles by observing the infrared radiation emitted in the 15-μm band from atmospheric carbon dioxide. The selective chopper radiometer on Nimbus 4 consists of six spectral filter radiometers. The four spectral channels observing the lower levels of the atmosphere are known as single gas cell channels. The optics of each channel includes a cantilever-mounted blade shutter that oscillates at 10 Hz and successively chops the field of view between the earth and a cold space source. The chopped radiation is then passed through a 10-cm path length of carbon dioxide cell, the cell pressure being set for each channel to define the viewing height in the atmosphere. Behind the CO_2 cell is a narrowband filter, the center of which is also different for each channel, and an infrared detector. For temperature sounding in the upper layers of the atmosphere, the other two channels operate on a slightly different principle and are known as double-cell systems. The technique consists of switching the radiation between two half-cells, semicircular in shape and of 1-cm path length, containing different pressures of carbon dioxide. The oscillating shutter used in the four other channels is replaced by a vibrating 45° mirror. During one half-period earth radiation passes through one half-cell and space radiation through the other. Table 4.16 shows spectral bands of the selective chopper radiometer and is applications.

A 16-channel selective chopper radiometer has been built for Nimbus 5 for better atmospheric measurements. The field of view is narrowed to 1.5°, compared to 10.0° for Nimbus 4. This change increases the chance of a clear field of view by making it possible to look through openings in the cloudy atmosphere. More spectral channels for better coverage and cirrus cloud measurement are included in the sensor design. Table 4.17 shows spectral applications of the Nimbus 5 selective chopper radiometer.

If a selective chopper radiometer is required to select atmospheric radiation from very close to the center of the emitted lines or high-altitude emission lines, then the gas cell pressure to be modulated is required. The radiation falling on the detector is modulated only at frequencies that lie within the absorption lines of the gas in the cell. The Nimbus 6 pressure

Table 4.16

Selective Chopper Radiometer Spectral Band and its Applications[a]

Channel	Spectral band center (cm)	Cell length (cm)	Cell pressure (atm)	Applications
1	668/775	1	0.01/0.05	Vertical
2	668/775	1	0.01/0.05	temperature
3	668/935	10	0.03/0.03	sounding
4	675	10	0.2	
5	697	10	0.6	
6	712	10	0.3	

[a] Courtesy of NOAA.

Table 4.17

Applications of the 16-Spectral-Band Selective Chopper Radiometer[a]

Spectral band (cm^{-1})	Application
668	Atmospheric vertical sounding,
689	0–18 km
707	
726	
668 (0.3[b]; 0.0[c])	Selective chopper channels for
668 (0.3; 0.03)	vertical temperature sounding,
668 (0.3; 0.1)	25–45 km
668 (0.3; 0.3)	
110	Cirrus cloud
202	detection
3710	
3805	
536	Water vapor emission
859	Atmospheric window: day/night
4550	Atmospheric window: nighttime only
2817	Atmospheric window: nighttime only

[a] Courtesy of NASA.
[b] Cell path length in centimeters.
[c] Cell pressure in atmospheres.

modulator radiometer (PMR) is a special sensor for measuring emission from CO_2 in the upper stratosphere and mesosphere (40–85 km). The PMR uses two methods of scanning the atmospheric emission lines from the height being modulated by the cell of CO_2 gas. These methods are pressure scanning and Doppler scanning.

In the pressure scanning technique the CO_2 cell pressure is sequenced by command to each of four values. Each pressure value changes the frequency at which cell transmission modulation occurs, and thus tunes the radiometer to a different part of the atmospheric lines. Since atmospheric radiation from different parts of emission lines originates at different altitudes, each mean cell pressure is sensitive to a different height. Thus, atmospheric temperature can be obtained at different altitudes.

If there is relative motion along the line of sight between the space radiometer and the emitting atmosphere, a Doppler shift occurs between the atmospheric emission lines and the absorption lines of the gas in the cell. Since the satellite speed is about 20 times the molecular speeds at normal atmospheric temperature, only about 5% of the satellite's velocity is required to produce a Doppler shift equal to the Doppler line width. By varying the Doppler shift, it is possible to scan the cell absorption lines across the atmospheric emission lines. The Nimbus 6 PMR instrument is designed so that its direction of FOV can be changed from nadir to $\pm 15°$ along the direction of the flight, thus introducing varying Doppler shifts, or the emission height can be tuned to a different altitude for atmospheric temperature measurements.

The Nimbus 6 PMR sensor consists of two similar radiometer channels, one with a cell 1 cm long containing CO_2 in the pressure range 0.5 to 3 mbar, and the other with a cell 6 cm long containing CO_2 in the pressure range 1 to 4 mbar. The first channel is intended to cover the 60–90-km region and the second to cover the 40–60-km region.

The stratospheric and mesospheric sounder (SAMS) sensor is the fourth in a series of multichannel infrared radiometers designed to measure emission of radiation from the upper atmosphere, where regular spectral filter-type radiometers do not give adequate performance. The SAMS sensor extends gas filter collection techniques to gases other than CO_2, in addition to viewing the limb direction of the atmosphere rather than employing vertical sounding as in the earlier selective chopper and pressure-modulated gas filter radiometers. The SAMS sensor is a 12-channel high-resolution infrared radiometer that measures atmospheric radiation from the earth's limb region. Global measurements are made of radiation from the molecular species CO_2, NO, CH_4, N_2O, H_2O, and CO.

The SAMS objectives are to determine the vertical temperature from atmospheric emission in the 15-μm CO_2 band from 15 to 80 km; to measure the vibrational temperature of the CO_2 band between 50 and 140 km; to measure vertical concentration profiles of CO, NO, CH_4, and H_2O from 15 to 60 km; and to monitor the vertical distribution of CO_2 and CO from 100 to 140 km and of H_2O from 60 to 100 km in order to study dissociation in the lower thermosphere. Figure 4.7 shows a schematic of the SAMS sensor. Figure 4.8 shows the sensor limb scan direction.

The improved stratospheric and mesospheric sounder (ISAMS) is an infrared gas filter radiometer that observes thermal emission and resonance fluorescence of solar radiation from the earth's atmospheric limb region. For vertical temperature measurements, the ISAMS spectral channels include the 15-μm band. For composition measurements, since the atmospheric temperature structure is known from measurements in the CO_2 band, values of effective emissivity can be deduced from the radiance measurements. With the help of models of atmospheric transmittance for the absorption bands of the various gases, values for the vertical concentrations of different gases may then be determined. The improved stratospheric and mesospheric sounder is a limb-viewing sensor for the upper atmosphere research satellite (UARS) program. Table 4.18 shows the sensor's spectral bands and applications.

The halogen occultation experiment (HALOE) employs satellite solar occultation observations to measure the vertical profiles of important atmospheric species in the upper atmosphere. These measurements include O_3, HCl, NO, NO_2, HNO_3, CH_4, H_2O, CO_2, HF, and other gases.

Table 4.18

ISAMS Spectral Bands and Applications[a]

Gas	Spectral band center (μm)	Applications
CO_2	14.2	Vertical temperature sounding
CO_2	15.0	15–80 km, 70–100 km
CO	4.6	15–60-km concentration profiles
NO	5.3	15–60-km concentration profiles
NO_2	6.3	15–50-km concentration profiles
H_2O	6.5	15–70-km concentration profiles
H_2O	30–100	15–70-km concentration profiles
CH_4	7.2	15–70-km concentration profiles
N_2O	7.9	15–60-km concentration profiles
O_3	9.8	15–60-km concentration profiles
HNO_3	11.4	15–45-km concentration profiles

[a] Courtesy of NASA.

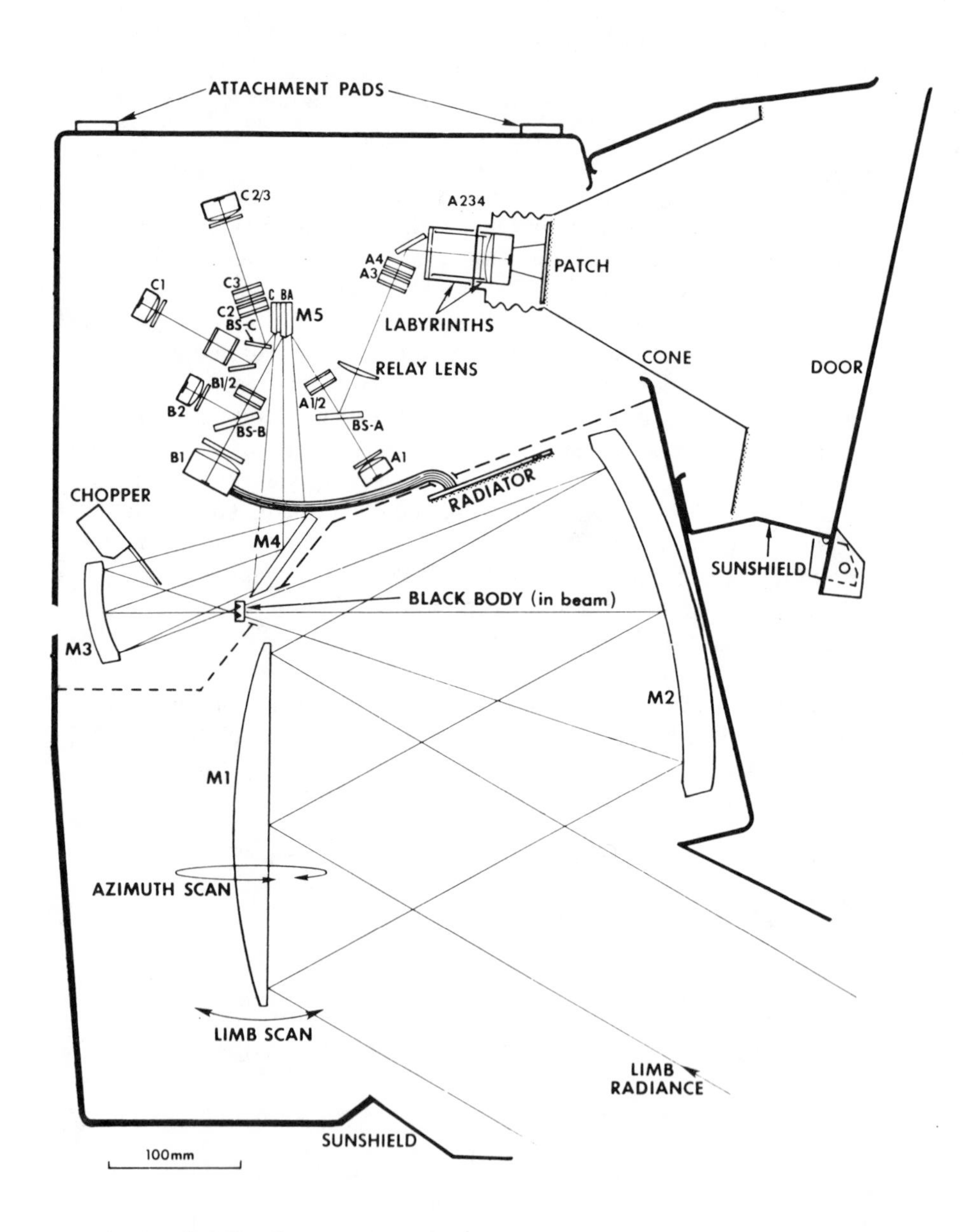

Fig. 4.7. SAMS radiometer schematic. (Courtesy of Hawker Siddley Co. and NASA.)

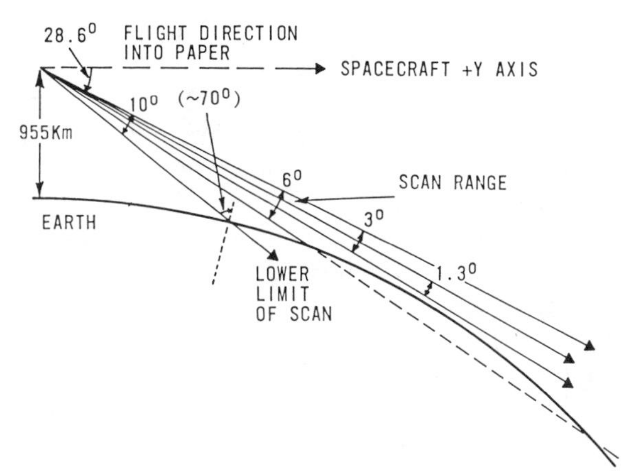

Fig. 4.8. SAMS sensor limb viewing angular relationship.

The HALOE sensor is an infrared sensor that uses gas filter correlation spectroscopy and broadband radiometry to measure vertical profiles of stratospheric trace gases. It has the ability to track the sun during occultation events. The sensor's spectral range is 2 to 12 μm with a resolution range of 20 to 200 cm^{-1}. Its detectors are InAs, HgCdTe, and a bolometer. The HALOE I sensor is the payload for ERBS and HALOE II is the payload for UARS.

The stratospheric aerosol measurement (SAM II) sensor provides vertical distributions of stratospheric aerosols in the polar regions of both hemispheres. It views a small portion of the sun through the earth's atmosphere during spacecraft sunrise and sunset in a single spectral band centered at 1.0 μm. The time-dependent radiance thus measured is combined with the spacecraft ephemeris data and local atmospheric density profiles, and then numerically inverted to yield a 1-km height-resolved vertical profile of aerosol extinction above the earth tangent point.

In the SAM II remote sensing application solar occultation by the earth's atmosphere is observed. The radiometer tracks the solar disk during each spacecraft sunrise and sunset event in order to produce an atmospheric extinction profile at the 1.0-μm wavelength from 10 to 40 km in vertical height. Retrieval of aerosol extinction profiles from the sensor data is accomplished through the following steps. First, the measured radiance data are reduced together with spacecraft ephemeris data into a single profile of limb optical thickness as a function of tangent height in the atmosphere. The high-altitude solar scan profiles are used as calibrated solar limb profiles in this process. Then the estimated Rayleigh extinction along each limb path is subtracted to obtain the aerosol extinction profile.

The limb radiance inversion radiometer (LRIR) on Nimbus 6 and the limb infrared monitor of the stratosphere (LIMS) radiometer on Nimbus 7 are designed to determine global vertical distributions of temperature and several important gases involved in the chemistry of ozone in the stratosphere. The LRIR measurements are made in each of four spectral regions: two in the 15-μm carbon dioxide band, one in the 9.6-μm ozone band, and one at 23–27 μm from 15 km in the lower stratosphere to 60 km in the lower mesosphere. The LIMS measurements are made in each of six spectral bands: one in the ozone 9.6-μm band, one in the nitrogen dioxide 6.3-μm band, one in the water vapor 6.2-μm band, one in the nitric acid 11.3-μm band, and two in the 15-μm band of carbon dioxide from 10 to 65 km.

The LIMS radiometer is a follow-on to the successful LRIR. The LIMS sensor is identical to the LRIR in many respects but is significantly different in that two detectors were added to the focal plane array. The sensor horizon scan rate was also decreased from 1° per second to $\frac{1}{4}$° per second to increase the sensor signal-to-noise ratio. A programmed scanning mirror in the radiometer scans the field of view of the sensor vertically across the earth's horizon. The scanned radiometric data are stored on magnetic tape for later transmission to the ground. The measured limb radiance profiles from the carbon dioxide bands are inverted by radiative transfer inversion algorithms to determine the vertical temperature distribution. This temperature profile and the radiance profiles in the gas channels are then used to determine the vertical distributions of atmospheric gaseous constituents.

The earth's limb atmospheric radiation is received by the sensor scan mirror and directed from the scan mirror to the telescope mirror system and sensor lens system. The focused radiation illuminates an array of HgCdTe detectors mounted on a cold finger embedded in the inner methane stage of a solid-cryogen cooler. The cooler outer stage is an ammonia tank.

Table 4.19 lists the LRIR and LIMS sensor spectral bands and FOV parameters.

The objective of the LRIR and LIMS sensors is to measure global variations of vertical atmospheric temperature, ozone, nitric acid, and nitrogen dioxide and to apply these measurements to the solution of upper-atmosphere radiation, chemistry, and dynamics. The LRIR and LIMS sensors use the passive infrared limb-scanning technique to measure the atmospheric parameters in the 10–65-km region. Figure 4.9 shows LRIR temperature retrieval from the limb measurements.

Table 4.19

LRIR and LIMS Sensor Spectral Bands and RFOV Parameters[a]

	LRIR		LIMS	
Gas	Spectral band (μm)	RFOV resolution (km)	Spectral band (μm)	RFOV resolution (km)
O_3	8.6–10.2	2 × 20	9.1–10.6	1.8 × 18
H_2O	23.0–27.0	2.5 × 25	6.5– 7.2	3.6 × 28
CO_2	14.4–16.9	2 × 20	13.5–16.8	1.8 × 18
CO_2	14.9–15.5	2 × 20	14.9–15.5	1.8 × 18
HNO_3			11.1–11.6	1.8 × 18
NO_2			6.2– 6.3	3.6 × 28

[a] Courtesy of NASA.

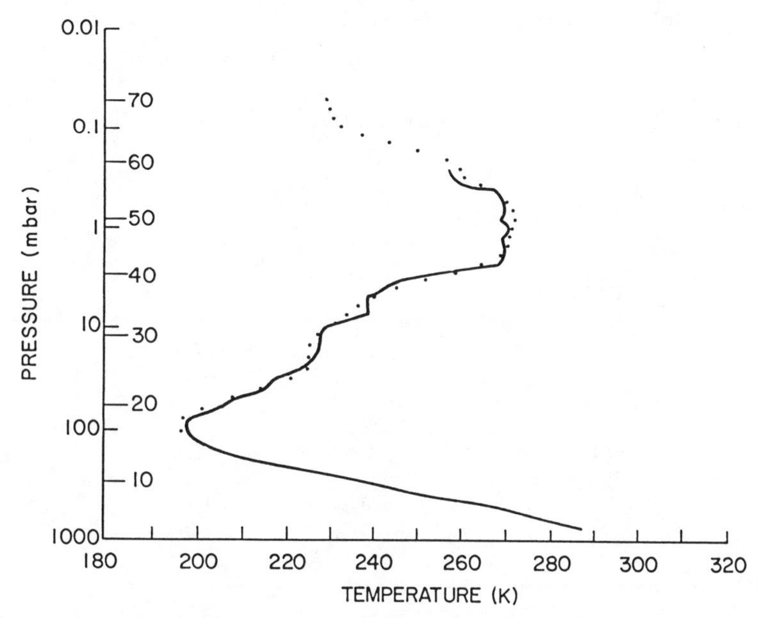

Fig. 4.9. LRIR temperature retrieval (dots) compared with rocket results (solid curve). (Courtesy of J. M. Russell and NASA.)

The temperature and humidity infrared radiometer (THIR) flown on Nimbus 4, Nimbus 5, Nimbus 6, and Nimbus 7 was designed to measure infrared earth radiation in two spectral bands centered at 6.7 μm and 11.5 μm. The 6.7-μm absorption channel provides information on the water vapor and cirrus cloud content of the upper troposphere and stratosphere, weather frontal systems, and jet streams. The 11.5-μm window channel provides temperatures of the land, ocean surfaces, and cloud tops and images of cloud cover and surfaces. The THIR sensor consists of an optical scanner module and an electronics module. The optical scanner provides the motion to produce cross-track scanning and sampling. It consists of a scan mirror, telescope, optical filters, beamsplitter, lenses, detectors, preamplifiers, power supply, and scan drive. The electronics module provides signal amplification and data processing capabilities, as well as the command input and housekeeping telemetry system for the spacecraft system.

The scan mirror, which is set at an angle of 45° to the scan axis, rotates at 48 rpm and scans in a plane perpendicular to the direction of the spacecraft motion. Radiation is reflected by the scan mirror to the telescope system and is separated by a beamsplitter into two infrared channels for detection and measurement. The earth radiation signals from the infrared detectors are coupled through a capacitor to the preamplifiers and transmitted to the electronics module for processing, conditioning, and filtering. Table 4.20 lists the THIR sensor parameters. Figure 4.10

Table 4.20

THIR Sensor Parameters

Sensor parameters	Channel 1	Channel 2
Aperture size, cm^2	110	110
IFOV, mrad	20	7
RFOV,[a] km	20	6.7
Spectral band, μm	6.5–7.0	10.5–12.5
Detector size, mm	0.67 × 0.67	0.22 × 0.22
Dwell time, msec	4.2	1.4
Bandwidth, Hz	115	345
NEI,[b] W/cm^2	4.35 × 10	3.0 × 10
NETD,[c] K	5.0 at 185 K	1.5 at 185 K
	0.26 at 300 K	0.28 at 330 K
S/N	3 at 185 K	19 at 185 K
	110 at 270 K	375 at 330 K

[a] Resolution field of view.
[b] Noise-equivalent irradiance.
[c] Noise-equivalent temperature difference.

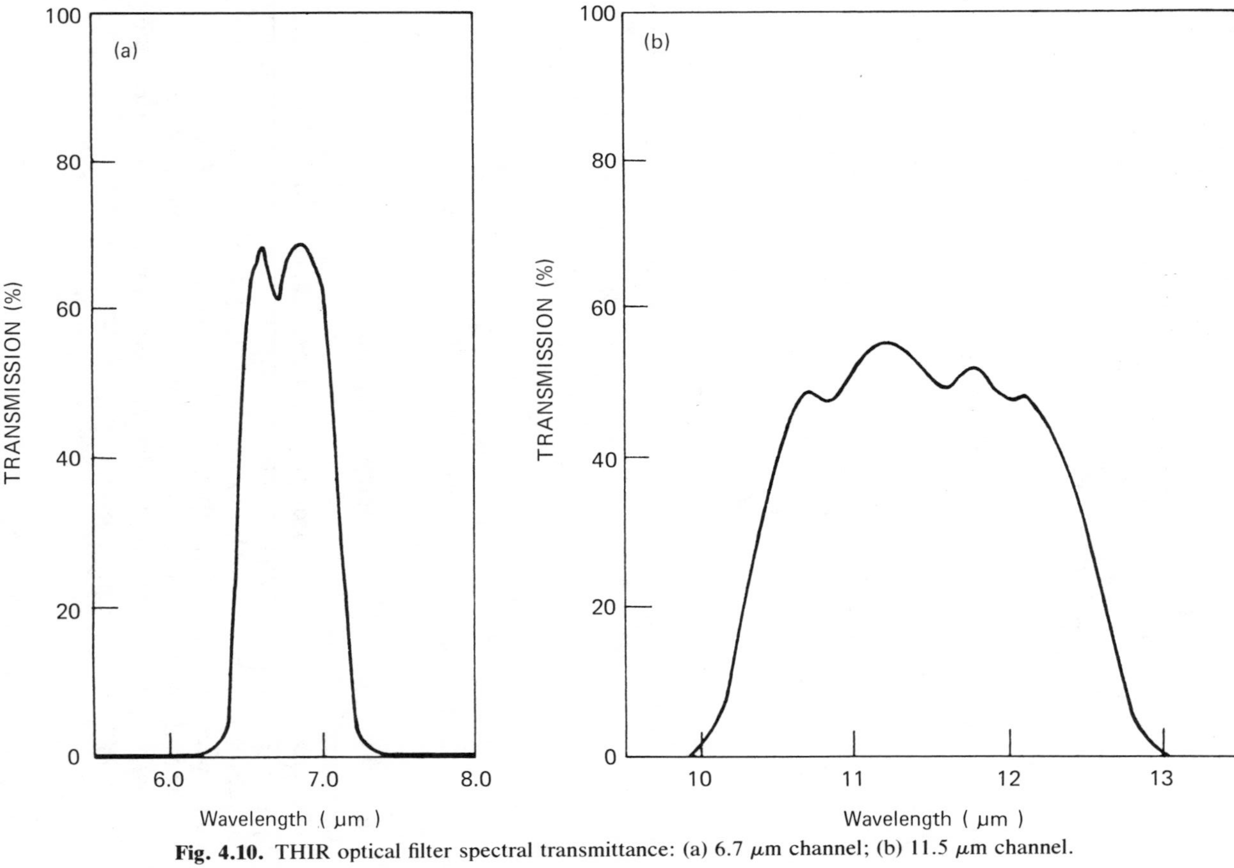

Fig. 4.10. THIR optical filter spectral transmittance: (a) 6.7 μm channel; (b) 11.5 μm channel.

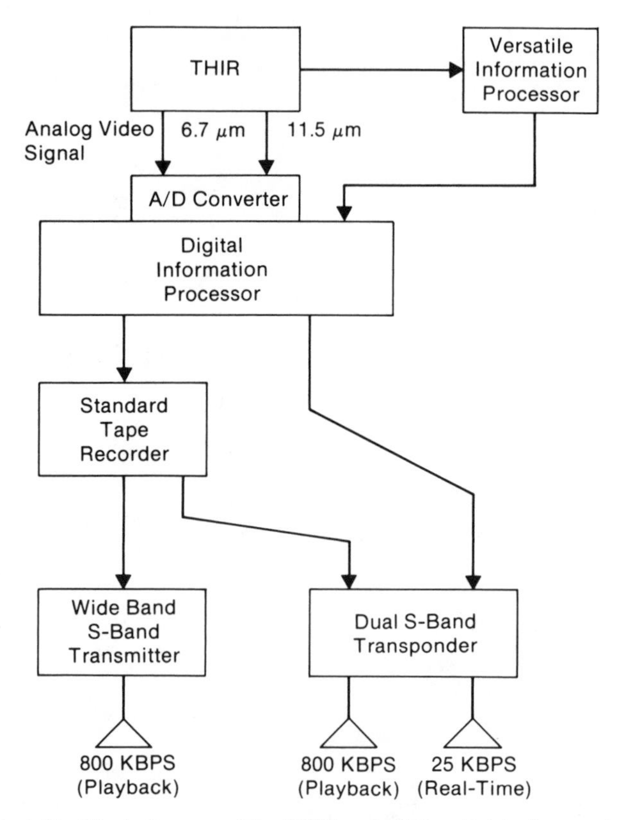

Fig. 4.11. Block diagram of the THIR and Nimbus 7 data flow system.

illustrates the optical filter spectral transmittance of the window band and the absorption band.

Figure 4.11 shows a block diagram of the THIR and Nimbus 7 data flow system. The window band and absorption band analog video signals are input to an analog-to-digital converter, which is part of the digital information processor. The housekeeping data are input to the versatile information processor, which multiplexes them with data from other Nimbus sensors and inputs them into the digital information processor. The composite data stream is then passed to one of three tape recorders or to the *S*-band transponder for direct transmission to the ground station.

4.5.1.2 *Land Applications*

The launch of Landsat 1 in 1972 opened a new era in land remote sensing as well as remote sensing for other fields. The satellite system offered a new vantage point, making new earth mapping data and remote sensing data available to a much wider community of earth scientists. The most immediate use made of the Landsat images of earth resources was in the identification and mapping of different areas of the earth's surface.

4.5.1.2.1 *The Multispectral Scanner (MSS).* The first multispectral scanner aboard Landsat 1 has returned 150,000 earth images with high information content since its launch in July 1972. The MSS sensor is designed to cover a 185-km swath width on the surface of the earth by angular motion of the object plane scan mirror. Each mirror scan generates six lines of data for each of the spectral bands and the subsequent scan, advanced by spacecraft orbital motion, generates another set of six scan lines. Thus, the oscillating scan mirror provides the cross-track scan while the orbital progress of the satellite provides the scan along the track.

The sensor primary and secondary mirrors are a Ritchey–Chretien telescope system. The focal plane system contains a matrix of fiber-optic elements that serve to dissect the image and carry the energy to separate detectors. The fibers are in a 4×6 array, corresponding to the four spectral bands and the six lines scanned per band. The square end of each single glass fiber forms a field stop of 86 μrad for 79-m ground resolution. The IR window band is 260 μrad for 240-m ground resolution. Two detectors cover adjacent lines during the scan to sweep out the same area as the six lines of the visible bands. The signal outputs of the 26 detectors are amplified, quantized, and multiplexed for transmission to ground stations. Table 4.21 is a summary of MSS sensor parameters. Figure 4.12 shows the MSS sensor schematic.

4.5.1.2.2 *The Thematic Mapper (TM).* The thematic mapper is a second-generation Landsat sensor system that provides improved information for earth resources remote sensing applications. The TM sensor is designed to achieve smaller imagery ground resolution, more separation between spectral band, and better calibration and radiometric accuracy. Landsat 4 and Landsat 5 operate in a circular sun-synchronous orbit at 705-km altitude and the TM scans a 185-km swath width. The repeat scanning cycle for a given swath occurs every 16 days at the same time each day, near 9:30 A.M.

A lightweight beryllium scan mirror is located directly in front of the telescope system. The telescope is an *f*/6 Ritchey–Chretien design, chosen to provide an image of a reasonable size at the focal plane. Directly beyond the telescope are the scan-line corrector, internal calibrator, and visible focal plane. The relay optics, near-infrared and infrared detector focal plane, and radiative cooler are located at the sensor end.

Table 4.21

MSS Sensor System Parameters

Parameter	MSS Nos. 4, 5	MSS Nos. 1, 2, 3
Orbit altitude, km	705.3	908.0
Repeater cycle, days	16.0	18.0
Telescope diameter, cm	22.9	22.9
f-number	3.6	3.6
Scan frequency, Hz	13.62	13.62
Scan angle, rad	0.260	0.200
IFOV, bands 1–4, μrad	117.2	86.0
RFOV, bands 1–4, m	82.65	78.6
Dwell time, bands 1–4, μsec	14.9	14.9
Filter 3-db point, bands 1–4, kHz	42.3	42.3
Nadir velocity, km/sec	6.75	6.46
Bandpass, μm		
Band 1	0.5–0.6	0.5–0.6
Band 2	0.6–0.7	0.6–0.7
Band 3	0.7–0.8	0.7–0.8
Band 4	0.8–1.1	0.8–1.1
Nominal number of samples per scan line	3270	3270

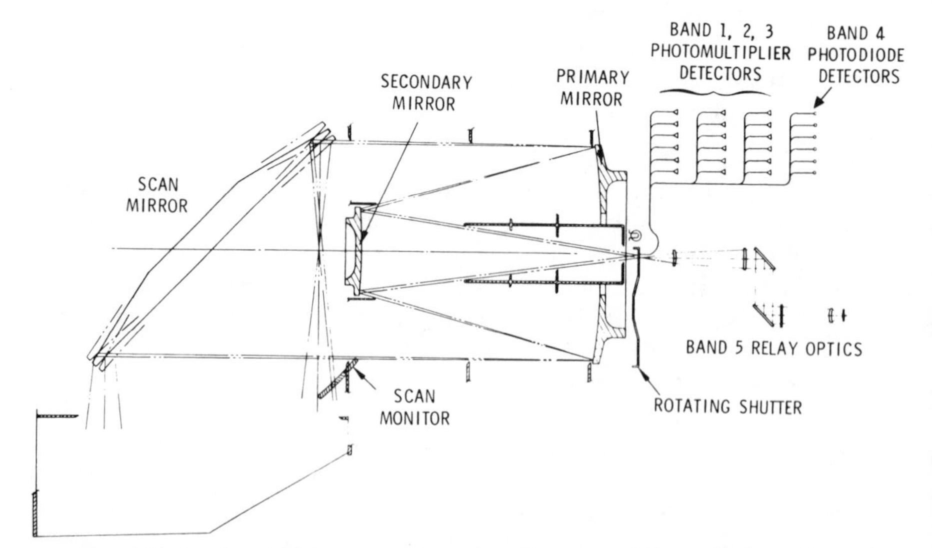

Fig. 4.12. Landsat MSS sensor schematic. (Courtesy of Santa Barbara Research Center.)

The primary and secondary mirrors are fabricated from ultra-low-expansion (ULE) glass, which has the desirable mirror qualities of fused silica and, in addition, exhibits a thermal expansion coefficient of essentially zero over a wide temperature range. The primary mirror has an aggregate structure, resulting in a weight reduction of approximately 60% over a comparable solid mirror blank.

The telescope tubular structure, which supports the primary mirror as well as all the other critical components in the optical system, including the radiative cooler for cooled focal plane detectors, is fabricated from graphite/epoxy composite materials. The graphite/epoxy structure has a very high stiffness-to-weight ratio and is an excellent material to match with the ULE glass mirrors.

The TM primary focal plane, which contains the square silicon detector arrays and critical preamplifier components for the first four spectral bands, is located at the primary focus location of the telescope. Two mirrors, one folding and the other spherical, are used to relay the image from the primary focal plane to the cooled secondary focal plane. The cooled focal plane contains InSb and HgCdTe detector arrays located in the radiative cooler at the image plane of the relay optics. The temperature range of the cooled focal plane is 90 to 105 K.

Figure 4.13 shows the MSS and TM detector focal plane formats. Figure 4.14 shows the TM sensor structure diagram, and Table 4.22 shows the TM system parameters.

4.5.1.2.3 *SPOT High-Resolution Visible (HRV) Sensors.* The French SPOT (Satellites Probatoires d'Observations de la Terre) satellite earth observation program will have a basic multimission platform design fitted with two identical high-resolution visible sensors. The HRV sensors have 20-m resolution in the multispectral band mode, while 10-m resolution will be possible with black-and-white imaging. The multispectral mode will use green, red, and near-infrared bands selected to fulfill several mission objectives, including good discrimination between different vegetation types and different soil types, a consistent relation between spectral reflectance and vegetational properties, and improved resolution for surface-water area and water penetration. Provision will be made for observation of areas vertically under the satellite, or at an oblique angle within a corrider 950-km wide, centered on the satellite ground track. For vertical viewing, each HRV sensor will observe a 60-km-wide swath. The satellite will pass over the equator at 10:30 A.M. local time.

Silicon CCD linear arrays have been used for the first-generation SPOT HRV sensor. For the panchromatic spectral band (0.5–0.7 μm) the detector comprises four standard CCD arrays of 1728 pixels (of which 1500 are

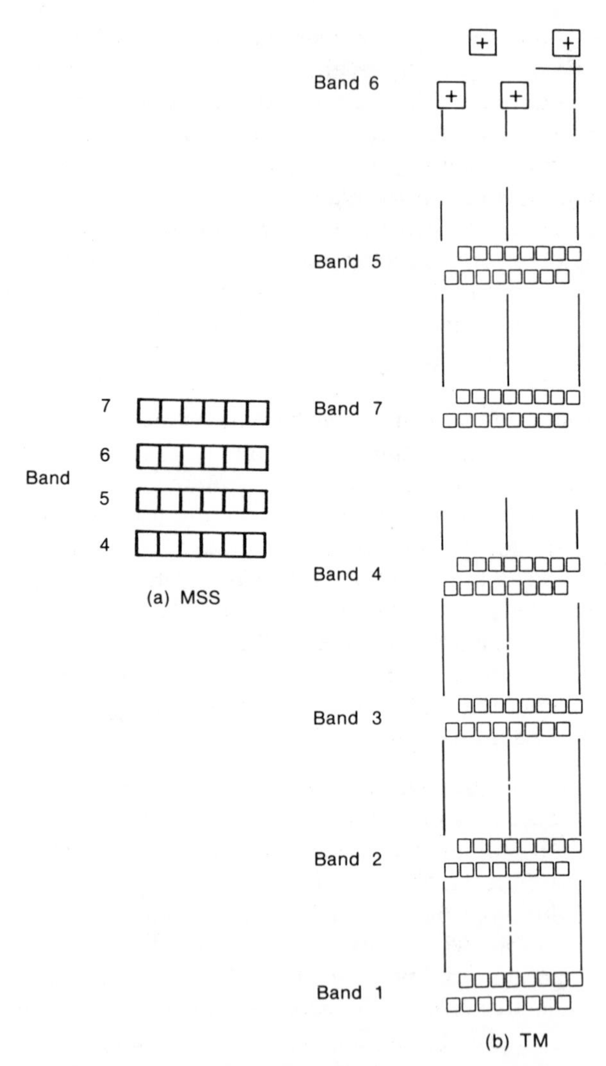

Fig. 4.13. MSS and TM sensor focal plane formats.

useful) and the linear array is optically aligned by a beamsplitter system. The 13-μm pixel size corresponds to 10 m on the ground, and 6000 pixels cover 60 km. The spectral bands centered at 0.55, 0.65, and 0.85 μm, with a near-0.1-μm half-width, use the same detectors but the pixels are grouped two by two. The 26-μm pixel size corresponds to 20-m ground resolution, and the linear array of 3000 pixels also covers 60 km.

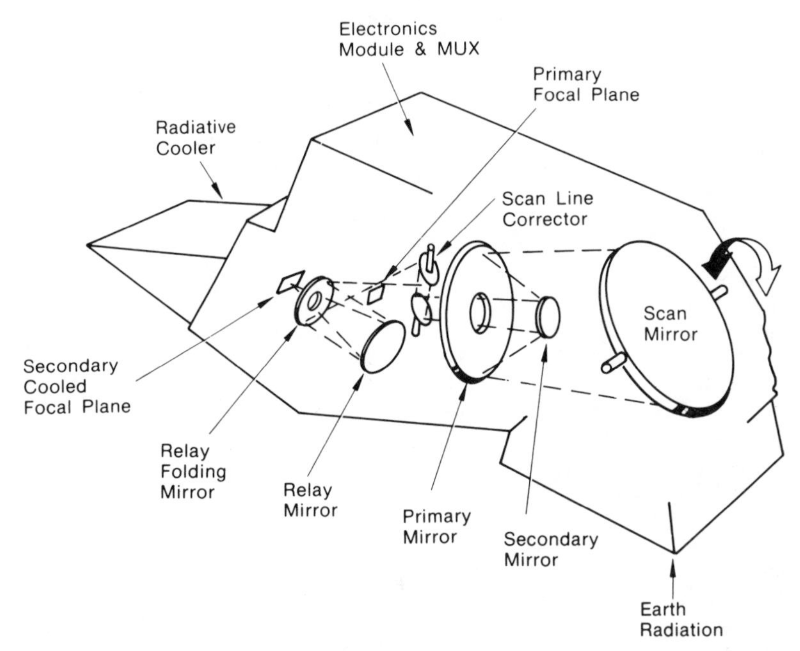

Fig. 4.14. TM sensor structure schematic. (Courtesy of Santa Barbara Research Center.)

Second-generation (in the 1990s) SPOT sensors will add 1.65- and 2.2-μm spectral bands for vegetation and geologic applications and 9- and 11-μm spectral channels for hydrology and thermal studies to offset atmospheric effects.

The SPOT HRV is a pushbroom space sensor for earth resources monitoring. Table 4.23 shows the SPOT HRV sensor system parameters.

4.5.1.2.4 *Multispectral Electric Self-Scanning Radiometer (MESSR).* The first Japanese earth observation satellite series, marine observation satellite (MOS), includes in its payload the multispectral electronic self-scanning radiometer (MESSR). The MESSR is designed to measure reflected image data from the earth's surface from a sun-synchronous orbit by pushbroom scanning over a swath width of 100 km. It has a linear CCD array and reads out image data to a serial time signal. The electronic scanning has an efficiency of almost 100%.

The distance between adjacent orbits of MOS is approximately 180 km. The sensor employs two optical systems, each covering an image swath width of 100 km with a CCD array of 2048 elements with a pixel size of 14×14 μm. This gives a ground resolution corresponding to 50×50 m.

Table 4.22

TM System Parameters

Orbital parameters	
Type	Sun-synchronous
Altitude	705.3 km
Inclination	98.21°
Eccentricity	0.001
Period	98.884 min
Repeat cycle	16 days
Nadir velocity	6.821 km/sec
Nadir angular velocity	9.671 mrad/sec
RFOV (visible and near IR)	30 m
RFOV (IR)	120 m
Optical/mechanical parameters	
Primary mirror area	1297.2 cm^2
Clear aperture area	1056.0 cm^2
Telescope focal length	243.8 cm
Scan mirror size	41.4 × 53.6 cm
Scan parameters	
Swath width	185 km
Active scan angle	0.1343 rad
Data rate	84.90 Mbps
Scan frequency	6.9967 Hz
Scan period	142.925 msec
Scan rate	4.42191 rad/sec
Scan efficiency	85%
Active scan time	60.743 msec
Turnaround time	10.719 msec
IFOV dwell time	9.611 μsec
Scan line length	6320 IFOV
Telescope parameters	
Primary diameter	40.6 cm
Primary area	1297.2 cm^2
Secondary obs. diameter	15.8 cm
Secondary obs. area	194.8 cm^2
EFL	243.8 cm
Primary reflectance	
Band 1	0.95
Bands 2–5	0.97
Band 6	0.98
Primary focal plane assembly	
Detector size	0.01036 cm^2
Detector sensitive area	107.4×10^{-6} cm^2
Center to center spacing in each row	0.0207 cm
IFOV size	42.5 μrad
Signal bandwidth	52.02 kHz

(continued)

Table 4.22 (*continued*)

Cold focal plane assembly	
Detector size	
Bands 5 and 7	0.00533 cm
Band 6	0.0207 cm
Detector sensitive area	
Bands 5 and 7	2.84×10^{-5} cm^2
Band 6	4.29×10^{-4} cm^2
Center-to-center spacing in each row	
Bands 5 and 7	0.0107 cm
Band 6	0.0414 cm
IFOV size	
Bands 5 and 7	43.75 μrad
Band 6	170.0 μrad
Focal plane temperature	90–105°K
Relay optics	
Folding mirror diameter	8.06 cm
Clear aperture diameter	7.44 cm
Second mirror diameter	14.22 cm
Clear aperture diameter	13.57 cm
Magnification	0.5
f-number	3.0
Spectral band range	
Band 1	0.45–0.52 μm
Band 2	0.52–0.60 μm
Band 3	0.63–0.69 μm
Band 4	0.76–0.90 μm
Band 5	1.55–1.75 μm
Band 6	10.4–12.5 μm
Band 7	2.08–2.35 μm
Sensor system weight	243 kg
Sensor power	332 W
Sensor envelope	0.66 × 1.1 × 2.0 m

Table 4.23

HRV Sensor System Parameters

Orbit	822 km, sun-synchronous, 60-km swath width
Spectral bands	
Band 1	0.50–0.59 μm
Band 2	0.61–0.69 μm
Band 3	0.79–0.90 μm
Band 4	0.50–0.90 μm
RFOV	
Bands 1–3	20 m
Band 4	10 m
Detector	6000 and 3000 CCD elements
Pixel size	13 × 13 μm, 26 × 26 μm.

Table 4.24

MESSR Sensor System Parameters

Orbit type	Sun-synchronous
Altitude	909 km
Period	103 min
Repeater cycle	17 days
RFOV	50 m
Spectral bands	
Band 1	0.51–0.59 μm
Band 2	0.61–0.69 μm
Band 3	0.72–0.80 μm
Band 4	0.80–1.10 μm
Swath width	200 km
CCD detector	2048 elements
Pixel size	14 × 14 μm

Each optical system provides an image of the earth's surface in four visible and near-infrared bands. The MESSR optics has adopted a dioptric lens system for wide-field-of-view applications. Table 4.24 shows the MESSR system parameters.

4.5.1.2.5 *Modular Optoelectronic Multispectral Scanner (MOMS).* The MOMS-1 is a payload of the Shuttle pallet satellite (SPAS) and is the first high-resolution pushbroom scanner to gather earth resources data from the Space Shuttle. It achieves 20-m ground resolution at 140-km swath width from the 300-km Shuttle orbit. The sensor uses a refractive lens systems to cover the total field of view to about ±35° with undistorted planar CCD focal planes. The dual-lens principle is applied for scan line extensions beyond one CCD array. The MOMS-1 uses 6912 CCD elements to generate image lines. There are four spectral bands for the second-generation MOMS sensor. The sensor could serve for earth monitoring applications, for example, for vegetation state measurement, soil and vegetation moisture measurements, and mineral and fossil resources exploration. Table 4.25 shows the MOMS sensor parameters.

4.5.1.2.6 *Multispectral Linear Array (MLA) Sensor.* The NASA multispectral linear array sensor is designed to provide for the evolutionary development of advanced sensor concepts and technologies and to provide a scientific basis for future remote sensing applications. The STS/MLA experiment is planned as a Space Shuttle observational research mission. It will collect data to support research in the science of the earth's surface and atmospheric science and to validate sensor system

Table 4.25

MOMS Sensor System Parameters

Orbit	300 km (Space Shuttle)
Spectral bands	
Band 1	0.6 ± 0.0025 μm
Band 2	0.9 ± 0.075 μm
Band 3	1.6 ± 0.1 μm
Band 4	2.2 ± 0.1 μm
CCD elements	6912 (4 × 1728)
RFOV	20 × 20 m
IFOV	67.5 μm
TFOV	26.2°
Data rate	40 Mbps

technology. The primary objective of this mission is to obtain new and valuable information on terrestrial land cover through the enhanced capabilities provided by the advanced multispectral linear array pushbroom technology. Off-nadir and stereoscopic data will also be obtained to investigate the physical properties of land cover classes over a wide range of atmospheric conditions. The MLA sensor system parameters are summarized in Table 4.26.

The reusable STS/MLA sensor consists of separate refractive optics for the visible/near-IR (VIS/NIR) and short-wavelength IR (SWIR) spectral bands, a solid argon cryogen cooler, and a fore/aft pointing mirror. The VIS/NIR linear array focal plane will incorporate two four-band (4 × 1024) butted silicon CCD detectors with integral spectral filters. The SWIR focal plane will incorporate the dual-band (2 × 512) Pd_2Si CCD array.

Figure 4.15 shows the STS/MLA bidirectional reflectance experiment.

4.5.2 *GEO Imaging and Sounding Applications*

Geostationary orbit is a unique vantage point from which to monitor the earth and its atmospheric systems. At an altitude of 35,800 km the period of the circular orbit is 24 hr. If the orbit lies in the plane of the earth's equator, the satellite and the earth turn through the same angular speed in the same time, so that the satellite is always above the same point on the equator. The GEO space sensor provided imaging of the earth's disk about every 20 min for environmental monitoring—so frequently that wind information can be extracted and storms can be detected. Infrared earth imaging can be obtained day and night. In 1966 and 1967, application

Table 4.26

STS/MLA Sensor System Parameters

Orbit altitude	250 km
Swath width	30 km
Sensor focal length	28.5 cm
Sensor aperture	8.14 cm
Spectral bands	
Band 1	0.46–0.47 μm
Band 2	0.56–0.58
Band 3	0.66–0.68
Band 4	0.87–0.89
Band 5	1.23–1.25
Band 6	1.54–1.56
CCD elements	
VIS/NIR	4 × 1024
SWIR	2 × 512
Pixel size	
VIS/NIR	15 × 15 μm
SWIR	30 × 30 μm
RFOV	
Bands 1–4	15 m
Bands 5 and 6	30 m
IFOV	
Bands 1–4	50 μrad
Bands 5 and 6	100 μrad
Data rate	40 Mbps
Solid cryogen cooler	2 W at 120°K
Weight	150 kg
Power	120 W

technology satellites ATS-1 and ATS-3 were launched into geostationary orbit above the Pacific and Atlantic Oceans. The GEO space sensor demonstrated the practicality of geostationary satellites for remote sensing of environmental data. Recognizing the merits of GEO the synchronous meteorological satellite (SMS) and geostationary operational environmental satellite (GOES) programs were developed for new GEO applications.

4.5.2.1 *The ATS Spin-Scan Cloud Camera (SSCC) Sensor*

The spin-scan cloud camera is a space radiometer consisting of a telescope and a photomultiplier tube at the focal plane of the telescope. The whole sensor can be tilted to a ±7.5° latitude elevation angle with a

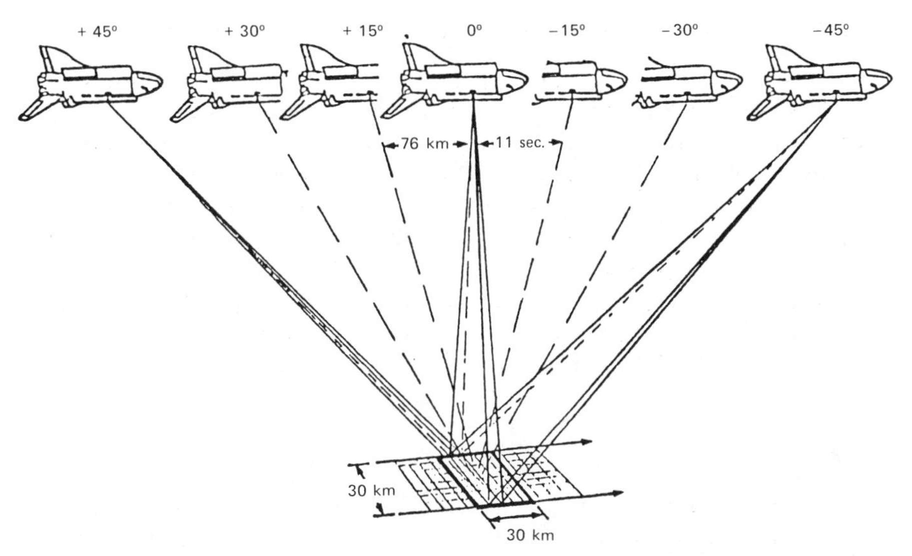

Fig. 4.15. STS/MLA multi-looking-angle and multispectral mission: off-nadir pointing for bidirectional reflectance measurements of terrestrial land cover. (Courtesy of NASA.)

precision step mechanism. The latitude step motion combined with the nominal 100-rpm spinning motion of the ATS provides coverage of the complete disk of the earth. The earth is scanned in about 2000 lines with a ground resolution of 4 km. The latitude step mechanism advances one 0.131-mrad step by command from the spacecraft on each revolution. This step occurs after the sensor has scanned past the earth. When the step mechanism has completed 2018 steps, which requires approximately 20 min, a limit switch initiates vertical retrace. The sensor returns to the north latitude limit position in a fast retrace mode, which takes 2 min. The scan-step process is then repeated for another picture of the earth's disk.

The sensor telescope is a Newtonian reflective system with an effective focal length of 25 cm, composed of a 12.5-cm-aperture fused quartz parabolic primary and a flat secondary mirror. The mirror surfaces are coated with aluminum and magnesium fluoride. The field stop at the focal plane is a pinhole providing an IFOV of 0.1 mrad. The earth radiation collected by the telescope and focused on the field stop aperture passes through it to a diverging lens and PMT. The photomultiplier tube is an end-looking type with an S-11 photocathode. The anode current, while the camera was scanning in space was well below 1 μA, and the power supply provided current at an adjustable 1935–2500 V dc.

4.5.2.2 *Multicolor Spin-Scan Cloud Camera (MSSCC) Sensor*

The MSSCC on ATS-3 was launched in November 1967 and positioned at 149°W over the Pacific Ocean. Several modifications were introduced in the second GEO sensor. The major change was a three-spectral-band design to improve discrimination of dim clouds from bright land, obtain better color contrast for both land and cloud features, and provide a longer visible wavelength channel with minimum scattering effects. The elevation scan angle was increased to 18° in 2400 scan lines for full earth coverage. Three PMTs were used in the design with fiber-optic leads from the focal plane aperture to the PMT window for signal detection.

Table 4.27 lists the sensor design parameters of the SSCC and the MSSCC.

4.5.2.3 *The Visible Infrared Spin-Scan Radiometer (VISSR)*

The VISSR is the main payload for the SMS and early GOES systems. It is the third type of spin scan space sensor for GEO applications. The VISSR is similar in operation to the ATS spin-scan sensors. Spacecraft spin generates the west–east scan and precision stepping of the scan mirror generates the north–south scan. The ground resolution in the visible has been increased fourfold (0.9 km), and an infrared band (9.0-km resolution) has been added for temperature measurements of clouds and the earth's surface, day–night cloud movement maps, and wind-field determination.

Table 4.27

ATS Sensor System Parameters

Parameters	SSCC	MSSCC
Telescope type	Newtonian	Wynn–Rosin
Primary mirror, cm	12.7	12.7 (ellipsoid)
Secondary mirror, cm	5.0	4.6 (spheroid)
IFOV, mrad	0.1	0.1
Spectral bands, μm	0.47–0.63	0.38–0.48 0.48–0.58 0.55–0.63
Photocathode	S-11	S-11 and S-20
IFOV dwell time, μsec	9.6	9.6
Scan lines per frame	2000	2407
Frame scan angle, degrees	15	18
Signal bandwidth, kHz	100	160
Weight, kg	9	10.5
Power, W	24 max	22 max

Earth radiation is collected by the VISSR scan mirror and reflected to the *f*/7.2 Ritchey–Chretien telescope with a 40.6-cm aperture diameter. At the primary focus, eight 0.06 × 0.06 mm fibers form the visible field stop aperture; the fibers transfer the radiation at the focal plane to the enhancement prisms on each of eight S-20 PMTs. Two germanium lenses relay IR radiation from the primary focus to two 0.13 × 0.13 mm HgCdTe detectors on the second-stage cold plate of the radiant cooler.

There are 1,812 IR image lines and 14,568 visible lines in the 20° × 20° full earth frame. The ground resolution at the satellite subpoint is about 0.9 km in the visible spectral band from 0.55 to 0.75 μm and 9 km in the 10.5–12.6-μm thermal window. The 20° frame time is 18.2 min and the retrace to the north starting position requires 1.82 min.

The HgCdTe detectors are cooled by a passive radiant cooler. The two-stage cooler has a demonstrated on-orbit capability of 70°K. Both cold stages have heaters to boil off any contaminants that may cryogenically pump out on the cold surfaces.

The VISSR scan mirror, primary and secondary mirrors, and sensor telescope housing are fabricated from beryllium. The SMS/GOES VISSR sensor is approximately 1.52 m long and 0.65 m in diameter. The sensor weighs only 65 kg. Maximum power is 25.5 W.

Figure 4.16 shows the GEO sensor focal plane formats. The geostationary meteorological satellite (GMS) and SMS/GOES VISSRs are iden-

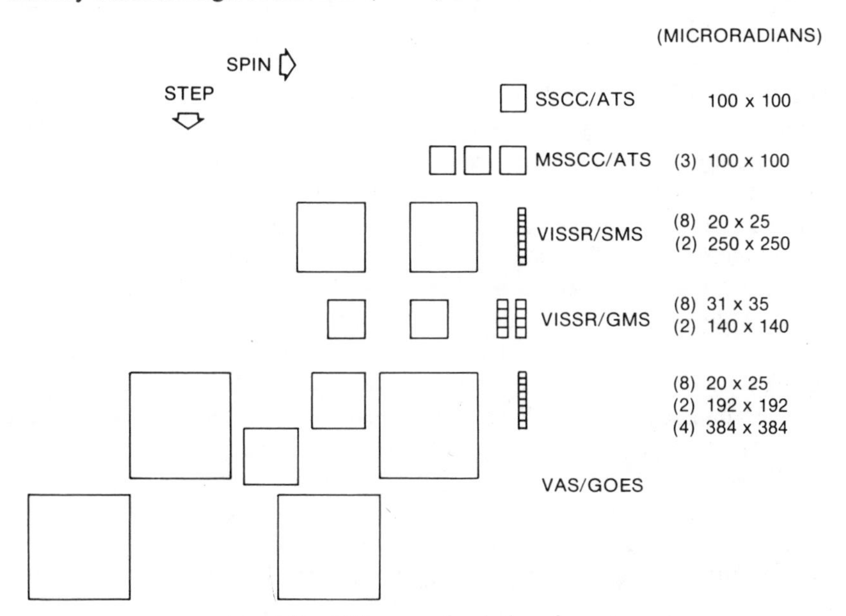

Fig. 4.16. GEO sensor focal plane formats.

tical in most respects except for the IFOV, as shown in Figure 4.16. At the subpoint the ground resolution of each visible line is 1.25 km and the IR resolution is 5 km. There is a redundant visible channel that can be switched into the downlink on command. The GMS VISSR and the spacecraft were built by Santa Barbara Research Center (SBRC) and Hughes for the National Aeronautics and Space Development Agency (NASDA) of Japan.

4.5.2.4 *VISSR Atmospheric Sounder*

From the GOES-4 spacecraft to the latest GOES, the environmental observation space sensor has been a new VISSR atmospheric sounder (VAS), which is a more sophisticated version of the VISSR. In its operational modes, the VAS scans the earth and gathers images in channels for reflected visible radiation and thermal IR radiation, as did the VISSR, and retains the VISSR imaging capability as an independent operational mode. The VAS has a new sounding capability from GEO. For multispectral scan imaging, the VAS is programmed to transmit signals from IR detectors in pairs: two small HgCdTe, two large HgCdTe, or two InSb detectors. By selective use of filters in combination with the detectors, a full frame of scans can image the earth in as many as three IR spectral bands.

The VAS is designed to operate in a spinning spacecraft of modest weight that can be launched to GEO. Design requirements for imaging and sounding capability dictate a telescope with relatively large optics. To avoid the great weight of high-precision optics made from conventional materials, such as glass mirrors mounted in Invar structures, the entire VAS structure and major optics are made of beryllium, a very light and stiff metal with the required dimensional stability.

The Ritchey–Chretien telescope system provides a larger field at the focal plane of the VAS, which is necessary to accommodate the IFOV. A relatively large focal plane is used in the VAS. Visible spectral signals are obtained at the principal focal plane. An optical fiber for each of the eight FOVs defines the field to be measured. Earth IR radiation is sensed by the HgCdTe and InSb detectors, which are cooled to 95°K to reduce detector noise. Because of the VAS sensor package design, the passive radiant cooler must be located away from the primary focal plane. This requirement is accommodated by relaying the IR signal to the secondary focal plane. The relay optics provide an appropriate location for an IR focal plane. Filter wheels are used to insert filters in the IR path only, and are used in multispectral scan imaging (MSI) and atmospheric sounding modes.

Table 4.28

VAS Sensor System Parameters

Telescope	Ritchey–Chretien
Sensor aperture	40.6 cm
Focal length	292.1 cm
IFOV	
Visible	20 × 25 μrad
IR	192 and 384 μrad
Detectors	
Visible	PMTs
IR	HgCdTe and InSb
Focal plane temperature	95°K (IR)
Spectral bands	0.6-μm window
	3.9-μm window
	11.1-μm window
	6.7-μm H_2O sounding
	7.2-μm H_2O sounding
	12.6-μm H_2O sounding
	4.44-μm temperature sounding
	4.52 μm temperature sounding
	13.33 μm temperature sounding
	14.01 μm temperature sounding
	14.25 μm temperature sounding
	14.48 μm temperature sounding
	14.73μm temperature sounding
Data rate	
Visible	1.74 Mbps
IR	0.52 Mbps
Size	1.5 × 0.65 m
Weight	75 kg

Table 4.28 shows the VAS sensor system parameters. Figure 4.17 shows the VAS sensor schematic.

4.5.2.5 *Meterosat Sensor System*

Meterosat was built for the European Space Agency for GEO remote sensing applications. The main payload of Meterosat is a radiometer, which provides information on cloud and surface IR temperatures and hence cloud height, while the visible channel shows cloud formation and enables wind velocity to be calculated with an accuracy of 3 m/sec. The water vapor distribution in the upper layers of the troposphere can also be provided by the radiometer output.

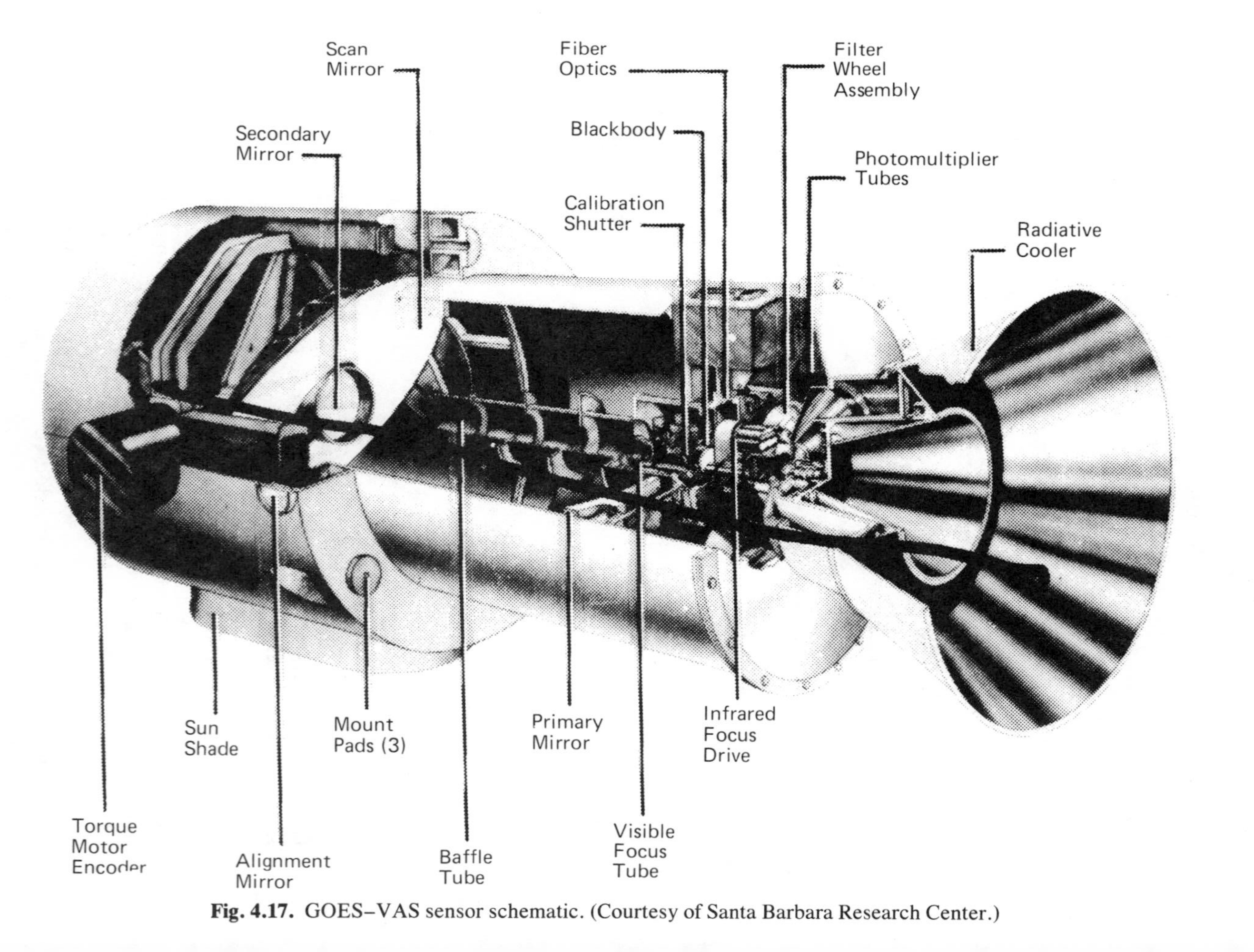

Fig. 4.17. GOES–VAS sensor schematic. (Courtesy of Santa Barbara Research Center.)

The Meterosat high-resolution radiometer is an electro-optical sensor that consists mainly of a Ritchey–Chretien telescope with a 40-cm aperture. Earth images are taken in the visible channel from 0.5 to 1.0 μm and in the infrared from 10.5 to 12.5 μm and 5.7 to 7.1 μm for window and water vapor remote sensing applications. From the GEO position, Meterosat observes the earth in a frame time of 25 min for one visible and one IR image. The ground resolution is 2.5 km in the visible and 5 km in the infrared. The visible images are composed of 5000 lines with 5000 sampling points each, and the infrared images of 2500 lines with 2500 sampling points each. Line scanning is done by rotation of the satellite, which spins at 100 rpm. Movement from one line to the next is done by step-by-step rotation of the telescope and detector system, synchronized with the satellite's rotation period.

Figures 4.18–4.22 show infrared pictures of the earth taken by the scanners from geostationary orbit at longitudes 135°W, 75°W, 0°, 55°E, and 140°E.

Fig. 4.18. Earth picture from GEO at 135°W. (Courtesy of NASA and NOAA.)

Fig. 4.19. Earth picture from GEO at 75°W. (Courtesy of NASA and NOAA.)

Fig. 4.20. Earth picture from GEO at 0°. (Courtesy of the European Space Agency.)

Fig. 4.21. Earth picture from GEO at 55°E. (Courtesy of NASA and NOAA.)

Fig. 4.22. Earth picture from GEO at 140°E. (Courtesy of the National Aeronautics and Space Development Agency, Japan.)

General References and Bibliography

Ando, K. J. (1983). The NASA MLA program. *IEEE Int. Geosci. Remote Sens. Symp.* **83CH1837,** 5.1–5.6.

Arking, A., and Vemury, S. (1984). The Nimbus 7 ERB data set: a critical analysis. *JGR, J. Geophys. Res.* **89**(D4), 5089–5098.

Blanchard, L. E., and Weinstein, O. (1980). Design challenge of the thematic mapper. *IEEE Trans. Geosci. Remote Sens.* **GE-18,** 146–160.

Chen, H. S. (1979). A polarization scanner for atmospheric radiation monitoring from geosynchronous satellite. *IEEE Space Instrum. Atmos. Observ. Conf.* pp. 27–29.

Curran, R. J., Kyle, H. L., Smith, B. J., and Clem, T. D. (1981). Multichannel scanning radiometer for remote sensing cloud physical parameters. *Rev. Sci. Instrum.* **52,** 1546–1555.

Drummond, J. R., Houghton, J. T., Peskett, G. D., Rodgers, C. D., Wale, M. J., Whitney, E. J., and Williamson, E. J. (1978). The stratospheric and mesospheric sounder (SAMS) experiment. *In* "The Nimbus 7 User's Guide," pp. 139–174. Goddard Space Flight Cent., Greenbelt, Maryland.

Engel, J. L., and Weinstein, O. (1983). The thematic mapper. *IEEE Trans. Geosci. Remote Sens.* **GE-21,** 258–264.

Gille, J. C., and Russell, J. M. (1984). The limb infrared monitor of the stratosphere: experiment description, performance, and result. *JGR, J. Geophys. Res.* **89**(D4), 5125–5140.

Gille, J. C., House, F. B., Craig, R. A., and Thomas, J. R. (1975). The limb radiance inversion radiometer (LRIR) experiment. *In* "The Nimbus 7 User's Guide," pp. 141–162. Goddard Space Flight Cent., Greenbelt, Maryland.

Houghton, J. T. (1975). The pressure modulator radiometer (PMR) experiment. *In* "The Nimbus 6 User's Guide," pp. 163–171. Goddard Space Flight Cent., Greenbelt, Maryland.

Hummer, R. F. (1975). Earth observation from geostationary orbit. *Proc. Soc. Photo-Opt. Instrum. Eng.* **62,** 103–113.

Ishizawa, Y., Kuramasu, R., Kuwano, R., and Nagura, R. (1980). Multispectral electronic self-scanning radiometer for MOS-1. *Proc. Int. Astronaut. Congr.* **31,** 1–11.

Jacobowitz, H., Soule, H. V., Kyle, H. L., House, F. B., and the NIMBUS 7 ERBE Team (1984). The earth radiation budget experiment: an overview. *JGR, J. Geophys. Res.* **89**(D4), 5021–5038.

Koenig, E. W. (1979). The TIROS-N high resolution infrared radiation sounder. *IEEE Space Instrum. Atmos. Observ. Conf.* pp. 61–67.

Lansing, J. C., and Cline, R. W. (1975). The four- and five-band multispectral scanners for Landsat. *Opt. Eng.* **14,** 312–323.

Lansing, J. C., Wise, T. D., and Harney, E. D. (1979). Thematic mapper design description and performance prediction. *Proc. Soc. Photo-Opt. Instrum. Eng.* **183,** 30–41.

Lennertz, D. (1975). The European meteorological programme and the Metrosat system. *Proc. Int. Astronaut. Congr.* **26,** 1–11.

McCormick, M. P., Mauldin, L. E., McMasters, L. R., Chu, W. P., and Swissler, T. (1978). Stratospheric aerosol measurement (SAM II) experiment. *In* "Nimbus 7 User's Guide," pp. 105–138. Goddard Space Flight Cent., Greenbelt, Maryland.

Russell, L. M., and Gille, J. C. (1978). The limb infrared monitor of the stratospheric (LIMS) experiment. *In* "The Nimbus 7 User's Guide," pp. 71–103. Goddard Space Flight Cent., Greenbelt, Maryland.

Slater, P. N. (1980). "Remote Sensing, Optics and Optical Systems." Addison-Wesley, Reading, Massachusetts.

Smith, W. L., Abel, P. G., Woolf, H. M., McCulloch, A. W., and Johnson, M. J. (1975). The high resolution infrared radiation sounder (HIRS) experiment. *In* "Nimbus 6 User's Guide," pp. 37–58. Goddard Space Flight Cent., Greenbelt, Maryland.

Smith, W. L., Hickey, J., Howell, H. B., Jacobowitz, H., Hilleary, D. T., and Drummond, A. J. (1977). Nimbus 6 earth radiation budget experiment. *Appl. Opt.* **16,** 306–318.

Stowe, L. L. (1983). Validation of Nimbus temperature humidity infrared radiometer estimates of cloud type and amount. *Adv. Space Res.* **2,** 15–19.

Suomi, V. E., and Krauss, R. K. (1978). The spin scan camera system: Geostationary meteorological satellite workhorse for a decade. *Opt. Eng.* **17,** 6–13.

Thompson, L. L. (1979). Remote sensing using solid-state array technology. *Photogramm. Eng. Remote Sens.* **45,** 47–55.

Tower, J. R., McCarthy, B. M., Pellon, L. E., Strong, R. T., Elabd, H., Cope, A. D., Hoffman, D. M., Kramer, W. M., and Longsderff, R. W. (1984). Visible and shortwave infrared focal plane for remote sensing instruments. *Proc. Soc. Photo-Opt. Instrum. Eng.* **481,** 24–33.

Yue, G. K., McCormick, M. P., and Chu, W. P. (1984). A comparative study of aerosol extinction measurements made by the SAM II and SAGE satellite experiments. *JGR, J. Geophys. Res.* **89**(D4), 5321–5327.

Chapter 5 | Passive Space Spectrometer Systems

5.1 Introduction

There are three ways to classify space spectrometer sensor systems. The spectrometer can be considered in terms of physical refraction, diffraction, and interference. The separation of earth radiation into its spectral bands can be accomplished by prism dispersion, grating diffraction, and radiation interference.

The satellite infrared spectrometer (SIRS), launched in April 1969 on Nimbus 3, was the first space grating spectrometer for vertical temperature sounding experiments with extended altitude coverage. Eight channels 5 cm^{-1} wide were located in the 15-μm band of CO_2, corresponding to weighting functions that cover the atmosphere from the ground to 35 km in height.

The first NASA space infrared interferometer spectrometer (IRIS) was also launched on Nimbus 3 for vertical temperature sounding and water vapor and ozone vertical concentration measurements. It measured the spectrum of the earth from 5 to 25 μm at a spectral resolution of 5 cm^{-1}. Nimbus 4 carried an improved version of IRIS with still higher resolution, 2.8 cm^{-1}.

The spectrometer resolving power is defined as $R = (\lambda/\Delta\lambda)$ for a spectral band from $-\lambda/2$ to $+\lambda/2$ centered at wavelength λ and is a dimensionless number. Table 5.1 shows the resolving power for different spectrometers.

The type of spectrometer that is chosen for space applications depends on several factors: the width of the earth's emission spectral lines, radi-

Table 5.1

Spectrometer Resolving Power

Spectrometer	Resolving
Prism	1,000–50,000
Grating	5,000–500,000
Interferometer	10,000–5,000,000

Table 5.2

Atmospheric Emission Line Broadening as a Function of Height

Altitude (km)	Linewidth (cm^{-1})
0	0.15
10	0.06
20	0.014
30	0.006
40	0.006

ance of the emission lines, and interference from other atmospheric components. Nothing is simpler than a prism spectrometer. The more complicated grating spectrometer provides better spectral resolution, and the still more complicated interferometer spectrometer is used for high-spectral-resolution atmospheric measurements.

The atmospheric emission linewidth is a function of temperature, pressure, and the density of the molecules. Thermal line broadening is illustrated by the linewidths versus altitude shown in Table 5.2.

5.2 Space Spectrometer Spectral Band Selection

5.2.1 *Climate Spectral Band Selection*

Backscattered solar ultraviolet radiation can be used to measure total ozone and its vertical distribution above the height of maximum ozone concentration. The Nimbus 7 solar backscatter ultraviolet spectrometer (SBUV) and total ozone mapping spectrometer (TOMS) are designed to measure extraterrestrial ultraviolet solar irradiance and the solar ultraviolet radiation backscattered from the earth and its atmosphere. Table 5.3 shows the ozone sounding spectral band selection.

5.2.2 *Environmental Spectral Band Selection*

High-resolution interferometers can be used to provide information on vertical profiles of temperature and gaseous components. The spectral band selection for limb measurements of atmospheric species is shown in Table 5.4.

Table 5.3

Ozone Sounding Spectral Band Selection

Wavelength (nm)	Applications
255.631	Total atmospheric ozone determination
273.587	Vertical ozone profile determination
283.078	UV solar irradiance measurement
287.681	
292.268	
297.565	
301.951	
305.851	
312.544	
317.540	
339.871	
312.465	Total ozone mapping
317.454	
331.195	
339.803	
359.904	
379.956	

Upper-atmosphere temperature and wind fields can be measured with a high-resolution interferometer in the oxygen and OH emission region at 80–350-km altitude. Table 5.5 lits the spectral bands used.

Present infrared temperature sounders are capable of retrieving vertical temperature with an accuracy of 2.5 K and vertical height resolution of 5–6 km and ground resolution of 17.4-km diameter. Current developments in space temperature sounders suggest that by 1995 the temperature accuracy will be 1 K and the vertical height resolution will be 2 km. The horizontal area resolution will be 5 × 5 km, which will reduce cloud interference. Four basic approaches can be considered for advanced sounder sensors: a set of interferometer spectrometers, a set of Fabry–Perot interferometer spectrometers, a large grating spectrometer, and a high-resolution gas filter spectrometer.

The spectral band selection for future temperature sounders is shown in Table 5.6.

5.2.3 *Ocean Sensor Spectral Band Selection*

The ocean space sensor is intended primarily for determining the content of ocean water—organic or inorganic particulate matter or dissolved

Table 5.4

Limb Measurement Spectral Band Selection

Component	Wave number (cm^{-1})	Spectral resolution (cm^{-1})	Limb view applications
HCl	2842–2844	0.1–0.4	Vertical concentration profile
O_3	2795–2805	0.1–0.4	
CO_2	2300–2390	0.1–0.4	
CO	2140–2170	0.1–0.4	
NO	1896–1899	0.1–0.4	
NO_2	1602–1608	0.1–0.4	
H_2O	1400–1560	0.1–0.4	
N_2O	1290–1293	0.1–0.4	
CH_4	1210–1250	0.1–0.4	
O_3	1000–1140	0.1–0.4	
Freon 12	920–940	0.1–0.4	
HNO_3	877–881	0.1–0.4	
ClO	832–837	0.1–0.4	
Freon 11	837–842	0.1–0.4	
CO_2	635–680	0.1–0.4	

substances. Ocean water which contains very little particulate matter, scatters as a Rayleigh scatterer with a deep bluish color. The ocean space sensor provides data for analyzing that color for the content of the water. The ocean is a poorly reflecting surface, and most of the radiance received by the space sensor is backscattered solar radiation from the atmosphere rather than reflected solar radiance from the ocean. The spectral bands of the ocean sensor coastal zone color scanner (CZCS) are listed in Table 5.7.

Table 5.5

Visible Emission Line Selection

Emission	Wavelength (nm)	Spectral resolution (nm)	Application
Oxygen line	557.7	0.1	Vertical temperature profile
Oxygen line	630.0	0.1	
Oxygen line	731.9	0.1	
OH	683.0	0.1	
	687.6	0.1	

Table 5.6

Temperature and Water Vapor Spectral Band Selection

Spectral band (cm^{-1})	Spectral resolution (cm^{-1})	Applications
2350–2400	0.1–0.3	Vertical temperature
630–670	0.1–0.3	profile
1710–1940	0.1–0.3	Vertical water vapor
		profile

5.3 Prism Spectrometer Design and Analysis

The prism spectrometer is an excellent space sensor candidate for land remote sensing applications. When a parallel beam of solar radiation reflected from the earth's surfaces passes through a prism, rays of different wavelengths are dispersed to varying degrees depending on the refractive index of the prism. The path of a ray through a prism of apical angle α is shown in Figure 5.1. The angle between the incident and refracted rays is the angle of deviation θ, which is equal to $\phi_1 + \phi_2 - \alpha$, while the sum of the angles of refraction is $\chi_1 + \chi_2 = \alpha$.

The resolving power is obtained from the equation

$$R = -B(dn/d\lambda) \tag{5.1}$$

where B is the length of the base of the prism and $dn/d\lambda$ is the dispersion. The equation shows that R does not depend on the prism apex angle and so does not depend on the length of the refracting side. Neither does the

Table 5.7

Ocean Sensor Spectral Band Selection

Spectral band (μm)	Applications
0.43–0.45	Chlorophyll absorption
0.51–0.53	Chlorophyll correlation
0.54–0.56	Yellow stuff
0.66–0.68	Chlorophyll correction
0.70–0.80	Ocean body delineation

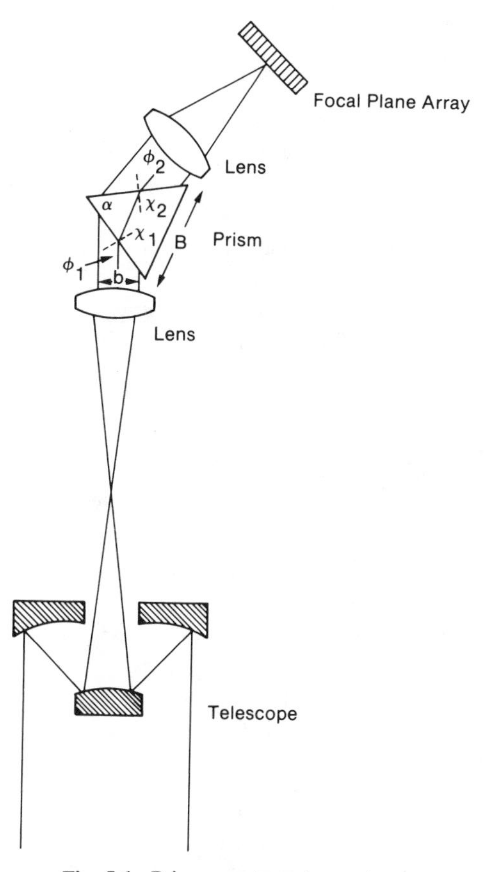

Fig. 5.1. Prism spectrometer system.

refractive index appear, but only its rate of change with wavelength. The smallest wavelength difference that can be resolved at 0.5 μm by a prism spectrometer having an equiangular silicate flint glass prism 5 cm on a side is calculated as follows:

$$dn/d\lambda = (1.650 - 1.615)/(0.6 - 0.4) \quad \mu\text{m} = 1.75 \times 10^3 \quad \text{cm}^{-1} \tag{5.2}$$

$$\begin{aligned} d\lambda &= \lambda/B(dn/d\lambda) = (5 \times 10^{-5})/5(1.75 \times 10^3) \\ &= 5.71 \times 10^{-9} \quad \text{cm} \end{aligned} \tag{5.3}$$

$$R = 5(1.75 \times 10^3) = 8750 \tag{5.4}$$

A prism spectrometer has some advantages over a grating spectrometer, such as absence of overlapping spectral orders and freedom from "ghosts," but the highest resolution possible with a prism spectrometer is an order of magnitude less than that possible with a grating spectrometer. There are two types of prism spectrometers. The Littrow-mounting type has been popular for many years; an off-axis paraboloidal mirror is needed for high-performance imaging applications. The Ebert mounting is superior; if a single-mirror system is used, the center of the mirror must be obscured to prevent transmission of the incoming beam via the center of the mirror to the exit slit. Figure 5.2 shows the Littrow prism spectrometer system.

The smallest spectral interval one would expect to resolve with a given combination of prism and mirror is determined by the following condition:

$$\Delta\lambda = (s/f)\, d\lambda/d\theta \tag{5.5}$$

where s is the slit width and f is the focal length.

It would be possible to resolve very closely spaced spectral lines by making the slit as narrow as is allowed by the sensitivity of a detector in

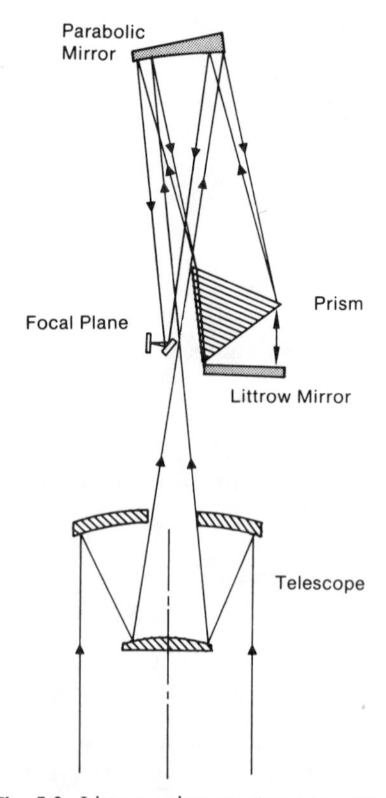

Fig. 5.2. Littrow prism spectrometer system.

which the resolution is energy-limited. For a diffraction-limited effect, the spectral resolution is expressed by

$$\Delta\lambda = (\lambda/b)\, d\lambda/d\theta \tag{5.6}$$

where b is the width of the radiation beam passing through the prism. From this equation one can see that making b large is the way to obtain high prism resolution. This is why a large prism is used in the sensor system for high spectral resolution.

5.4 Grating Spectrometer Design and Analysis

5.4.1 *The Slit Grating Spectrometer*

A grating spectrometer for atmospheric sounding consists of an optically flat blank coated with an aluminum or gold film in which a great number of accurately parallel and equidistant grooves are ruled. The density of grooves may range from a few hundred to a few thousand per inch, depending on the spectral range to be covered. There are two types of gratings. In a reflection grating, the light is reflected from close parallel grooves, or rulings, for the dispersion of earth radiation. In a transmission grating, the light passes through close parallel slits for the dispersion of earth radiation. The plane transmission grating shows such poor performance compared with the reflection grating system that plane reflection gratings are used for high-performance space spectrometers.

The resolving power of a grating system can be written as

$$R = mN \tag{5.7}$$

Thus it depends only on the product of the diffraction order m and the number of lines N in the ruled surface, and not on the wavelength or grating spacing. The main advantage of gratings is in the infrared region from 2.0 to 20 μm, where they offer the good resolution that is required in remote sensing and atmospheric sounding applications. Most large gratings do not attain a maximum resolving power of more than 300,000.

Grating mountings can be classified in two groups, each with a number of variants to suit particular needs. The Littrow mounting employs a single concave mirror, which acts as both collimator and focusing element. Normally this mirror should be an off-axis paraboloid, but a spherical mirror can be used at the cost of some extra aberration. The chief advantage of this mounting is its compactness. Its optical performance is comparatively poor if it is used over a range of grating angles greater than a few degrees. The Ebert mounting is the most important one for the construction of high-performance grating spectrometers. It was first described by H. Ebert in 1889 and subsequently rediscovered by W. Fastie, who described it in 1952, and for this reason it is often known as the Ebert–Fastie mounting. The basic difference between it and the Littrow mounting is that with the Ebert mounting two different optical elements or two different areas of the same element are used to collimate and focus the earth radiation. With an Ebert mounting the spectrometer has greater resolving power and uses long slits to improve the throughput.

The chief variants on the Ebert system are (a) the Czerny–Turner mounting, where two separate spherical mirrors are used side by side and the slits are placed on either side of the grating; (b) the crossed Czerny–Turner mounting, which suppresses scattered light by means of internal baffles and prevents the entrance slit from seeing the second mirror, or the exit slit from seeing the first; and (c) the Newtonian Ebert mounting, which reduces off-axis aberrations and can be used to remove spherical aberration for the highest possible spectral resolution.

The basic Littrow grating spectrometer system is shown in Figure 5.3.

Like the prism spectrometer, the plane grating spectrometer requires the use of lenses or mirrors to collimate and focus the incident and emergent radiation. H. Rowland investigated whether a grating could be ruled on a concave mirror, and the concave mirror used to focus the spectra formed by the grating. This reduces the spectrometer to three subsystems: a slit, a concave grating, and a focal plane. The concave grating spectrometer obeys the same equation for diffracted images as does the plane grating spectrometer:

$$m\lambda = d(\sin \alpha + \sin \beta) \tag{5.8}$$

where m is the grating order, λ the wavelength of the diffracted wave, d the grating spacing, α the angle of incidence, and β the angle of diffraction.

Order $m = 0$ corresponds to specular reflection; $m = 1, 2, 3, \ldots$ give diffractions of the first, second, third, . . . orders. The grating grooves can be blazed for a particular diffraction order and wavelength. To cover a wide range of wavelengths, the same grating may be used in several successive orders. Thus a grating having 2500 lines per inch may be used in the first order for the wavelength range 5–15 μm, in the second order for 3.5–5 μm, in the third order for 2.5–3.75 μm, and in the fourth order for 2–2.25 μm. Figure 5.4 illustrates the angular relation between the incident and diffracted rays.

Most gratings are used in the first order to avoid overlapping and obtain high efficiency over a wide spectral range. However, many systems oper-

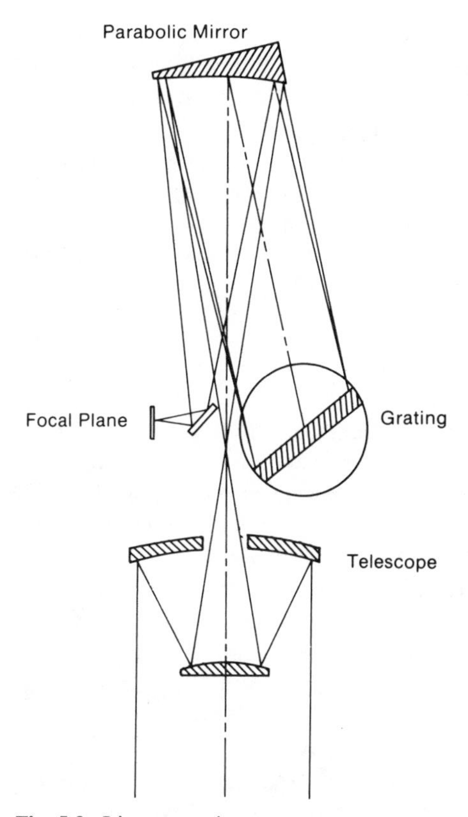

Fig. 5.3. Littrow grating spectrometer system.

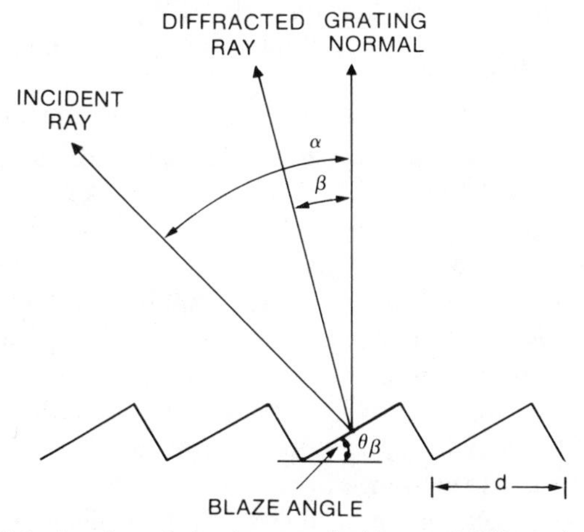

Fig. 5.4. Angular relation between incident and diffracted rays.

ate successfully in the second or even higher orders. In echelle gratings, orders below 5 are rarely used. Thus some kind of filtering is needed to prevent order overlapping, unless the system has a laser source. An echelle grating looks like a coarse grating, except that it is used at such a high angle that the steep side of the groove becomes the optically active facet. Typical echelle groove spacings are 316, 79, and 31.6 lines/mm, all blazed at 63.26°; high resolution can be obtained by using an echelle 254 mm wide. Echelles are used in high orders, from 5 to over 100, so that order sorting is necessary in the system design.

5.4.2 *The Grille Grating Spectrometer*

The grille spectrometer is basically a slit spectrometer in which the entrance and exit slits are replaced by a two-dimensional pattern of alternatively transparent and opaque zones, or a grille. The exit grille is a monochromatic image of the entrance grille through the sensor. The entrance and exit slits of a slit spectrometer limit the radiation that reaches the focal plane detectors, and consequently limit the signal output. The gain of a grille spectrometer is generally 100 to 300 in comparison with a slit spectrometer. The grille spectrometer sensor comprises a dispersive system and an entrance and exit collimator. Its mode of operation is also about the same as that of a slit spectrometer; that is, the spectra are obtained directly from the sensor. The entrance grille has a usable area of nearly 3 × 3 cm obtained by photoengraving material supported by an optical plate, depending on the spectral region of interest. The spectral range is 1–30 μm and the resolving power is 30,000 at about 3 μm. For an extended atmospheric weak emission source, the grille spectrometer has an advantage over the slit spectrometer in signal-to-noise ratio performance.

The structure of the grille is such that the opaque and transparent zones are limited by equilateral hyperbolas having the general equations $y = N/kx$, where N is an integer. The x and y axes are the common asymptotes of all the hyperbolas of the grille. This family of curves cuts off equal segments varying inversely with the distance of the line from the asymptote. The number of zones is nearly 600 hyperbolas on a 3 × 3 cm square. Each gap is 0.05 mm wide, and a slit of equivalent resolving power would be 0.07 mm wide.

Images of the entrance grille in different wavelengths will be spaced across the exit focal plane, and only for one wavelength will there be exact superposition of the image of the entrance grille on the exit grille. For a negative exit grille, none of the radiation of this wavelength passes through the grille. With the association of these positive and negative grille systems, the difference between the radiation carried by these two beams is a measure of the spectral interval of interest. In order to obtain these two beams, the opaque surface of the grille is made reflective, which complements the beam that is transmitted through the same grille. The difference between the transmitted radiation and the reflected radiation is measured by the detectors with an ac amplifier that is tuned to the frequency of the two beams. Thus, selective modulation of a narrow spectral domain about the earth radiation is obtained. In order for this signal difference to permit a precise radiation measurement, the optical path of the two beams must be made perfectly symmetrical, which requires very delicate adjustments. For convenience, another operating system as an alternating system. It can be seen that alternating movement of the image of the entrance grille parallel to the direction of the slit of a grating spectrometer system produces selective modulation of a single spectral element. The alternation of the image is obtained by vibrating the collimator's parabolic mirror about its horizontal axis. The sensor signal is the difference between horizontal asymptotes with merged and shifted grilles. This method of operation gives only half the radiation of the previous method. However, a single grille can be used, in reflection at the entrance and in transmission at the exit or vice versa. The center of symmetry of the grille is at the focus of the collimator mirror of the Littrow mounting. This alternating system is very simple compared with the previous method.

5.5 Interferometer Spectrometer Design and Analysis

5.5.1 *Fabry–Perot Interferometer Spectrometer*

A spectral scanning Fabry–Perot interferometer spectrometer can be designed by providing means for changing the wavelength of the atmospheric radiation passing through the étalon and for measuring and recording the radiation as a function of wavelength with the detectors at the focal plane. The variables that can be altered to effect the spectral scan are n, the refractive index, l, the spacing between the étalon plates, and θ, the angle of incidence with respect to the surface of the plate.

In pressure scanning l and θ are held fixed and the refractive index n of the gas between the plates is changed by changing the gas pressure in a windowed gastight cell inside the Fabry–Perot interferometer. The scanning spectral range that can be achieved by this method is

$$\Delta\nu = \nu \, \Delta n \tag{5.9}$$

or

$$\Delta\lambda = \lambda\ \Delta n \tag{5.10}$$

where Δn is the maximum change in index, ν the wave number, and λ the wavelength.

For air at standard conditions $n = 1.00029$ in the visible and the spectral scan range per atmosphere is about $\pm 5.0\ \text{cm}^{-1}$ at $17{,}330\ \text{cm}^{-1}$ or ± 1.25 Å at 5770 Å.

Mechanical spectral scanning can be achieved when the plates are mounted with a provision for carefully changing the plate spacing l in a linear uniform manner. The scan is dependent on l at fixed n and θ. The mechanical scan is related to the change dl in l through the equations

$$\Delta\nu = -\nu\ dl/l \tag{5.11}$$

$$\Delta\lambda = \lambda\ dl/l \tag{5.12}$$

For mechanical scanning the range increases with decreasing l, whereas for pressure scanning the total spectral range scanned is independent of l.

Angular spectral scanning can be achieved by virtue of a change in the angle θ with respect to the optical axis. The spectrum is scanned as the angle θ is changed simply by tilting the étalon.

Figure 5.5 shows a diagram of the Fabry–Perot spectrometer system.

The Fabry–Perot étalon usually consists of two parallel disks of optical glass material, with the inner surfaces polished over the whole surface area to better than $\frac{1}{50}$ or sometimes $\frac{1}{100}$ of a wavelength of HeNe laser light. These two flat surfaces are coated with special coating so that most of the radiation incident upon them is reflected and only a small fraction is transmitted. The Fabry–Perot étalon disperses the radiation passing through it in two dimensions, both perpendicular to the optical axis. The Fabry–Perot étalon forms a spectral ring system when radiation from the earth's atmosphere is transmitted. The rings can be projected onto a focal plane by a lens system. Different wavelengths produce rings of different

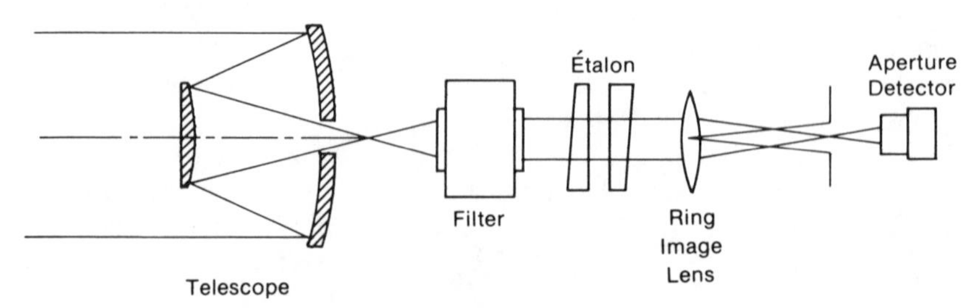

Fig. 5.5. Fabry–Perot interferometer spectrometer.

radius. Thus the dispersion is radial in two dimensions, and by cutting an aperture opening field stop in the focal plane certain wavelengths can be detected while all others are rejected. This aperture performs the same function as the exit slit in a prism or grating spectrometer. The Fabry–Perot interferometer requires no entrance slit; thus the throughput is much better than that of a grating or a prism spectrometer. The size of the aperture determines the resolution and the resolving power of the sensor. The angular diameter of the aperture can be written as

$$\theta^2 = 8/R \tag{5.13}$$

where θ is the angular diameter of the hole and R is the resolving power.

The aperture stop size is given by

$$d = 2f(2/R)^{1/2} \tag{5.14}$$

where d is the aperture size and f is the focal length of the lens system.

5.5.2 *Michelson Interferometer Spectrometer*

The Michelson interferometer spectrometer is a type of spectrometer in which the detector simultaneously receives radiance signals from different parts of the spectrum through multiplex transmission of the radiance signals. The Michelson interferometer is shown schematically in Figure 5.6. It consists of two radiation reflector mirror systems, one of which, M_1, is stationary, while the other, M_2, moves at a constant velocity v cm/sec. There is a beamsplitter system at which the incident radiation beam is divided and later recombined after a path difference has been introduced between the two radiation beams by the two mirror systems.

If there is a laser beam input to the Michelson interferometer and the moving mirror of the interferometer is adjusted so that the optical path difference is zero, the two laser beams will be in phase and the interference beam formed by the beamsplitter will appear bright. If the mirror is moved mechanically one-fourth of the laser wavelength or one-half of the wavelength of the optical path, then the interference beam will be 180° out of phase and the beam will appear dark. If the mirror is moved continuously the interference beam at the focal plane or at the detector will change from bright to dark or maximum to minimum signal output; the equation of the output can be written as

$$I(x) = B(\nu)\cos(2\pi x\nu) \tag{5.15}$$

where $I(x)$ is the signal output or the interferogram, $B(\nu)$ the intensity of the laser source at wave number ν, and x the distance of the mirror movement.

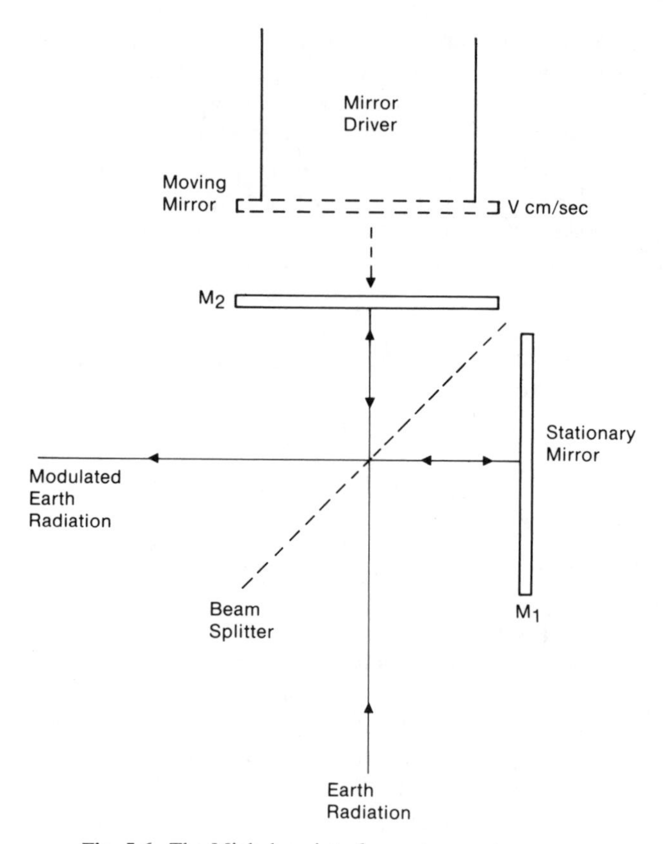

Fig. 5.6. The Michelson interferometer spectrometer.

The oscillation of the signal output depends on two parameters: the wavelength of the radiation and the velocity of the moving mirror. Hence the interferogram contains very high optical frequency signal information in the form of an oscillation.

For broadband spectral radiation, the interferogram will be the summation of all the oscillations caused by all the wavelengths in the earth radiation source. At zero optical path difference between the two mirrors, all of the radiation is in phase and there is a maximum interference signal output. As the moving mirror is moved away from the zero optical path position, wavelengths of different amplitude and phase add to each other with smaller maxima and minima. The radiation interference signal output that is caused by the moving mirror at each sampling position is called a broadband spectral radiation interferogram, as shown in Figure 5.7. The interferogram can be expressed by the following equation:

$$I(x) = \int_{-\infty}^{+\infty} B(\nu) \cos(2\pi x\nu)\, d\nu \tag{5.16}$$

The earth radiation spectrum can be found by taking the Fourier transform of the interferogram as shown in the following equation:

$$B(\nu) = \int_{-\infty}^{+\infty} I(x) \cos(2\pi x\nu)\, dx \tag{5.17}$$

For an asymmetric interferogram, the interferogram and the radiation spectrum must be treated by a complex Fourier transformation. By performing a Fourier transformation on the earth radiation interferogram from the time or distance domain to the frequency domain, the earth radiation spectrum can be obtained for remote sensing applications.

For a constant mirror velocity, there is a liner relation between the wave number of the incoming earth radiation and the frequency of the detector signal, as shown in the following equation:

$$f = 2V\nu \quad \text{Hz} \tag{5.18}$$

where f is the frequency of the detector signal, V the velocity of the moving mirror, and ν is the wave number of the earth radiation. For example, with a mirror velocity of 5 cm/sec, earth radiation at 10-μm wavelength (1000 cm^{-1}) will produce a detector signal of 5000 Hz.

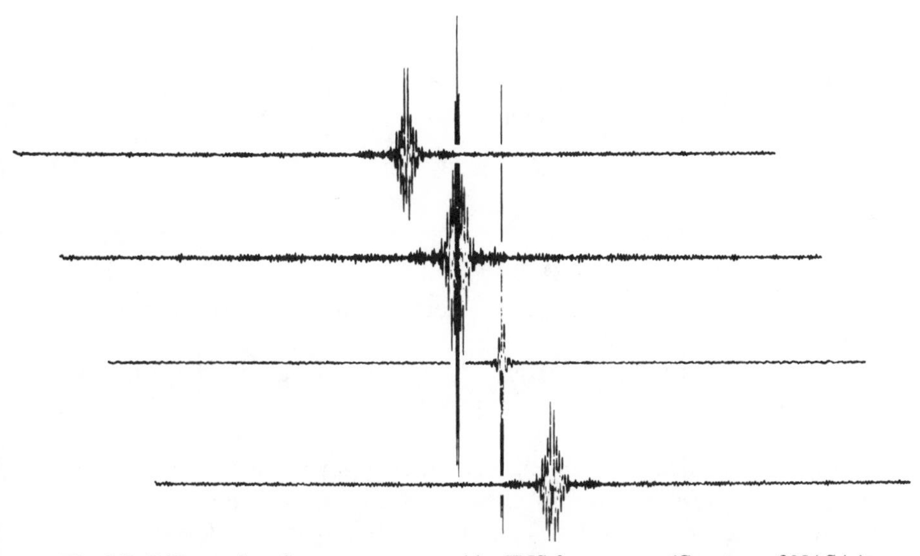

Fig. 5.7. Different interferograms measured by IRIS from space. (Courtesy of NASA.)

The relations between the parameters of the interferometer spectrometer system are listed in the following equations:

$$\Delta\nu = 1/2d \tag{5.19}$$

$$V = 1/4T(\Delta\nu) \tag{5.20}$$

$$f_s = 2\nu_s V \tag{5.21}$$

$$\Delta f = 2\nu_r V \tag{5.22}$$

$$f_0 = 2\nu_0 V \tag{5.23}$$

where $\Delta\nu$ is the spectral resolution, d the maximum optical path difference, V the moving mirror velocity, T the interferogram sampling time, f_s the sampling frequency, Δf the signal bandwidth, f_0 the signal central frequency, ν_s the sampling density in wave numbers, ν_r the free spectral range, and ν_0 the central wave number of the spectral range.

Usually, three interferometers form a complete interferometer spectrometer. These are the main Michelson interferometer for the earth radiation spectrum, the laser interferometer for sampling the interferogram, and the white-light interferometer for fringe-counting zero reset use. The laser interferometer produces the fringe-referenced sampling frequency for the interferometer spectrometer, which can be calculated as $\nu_L = 1/\lambda = 1/0.6328867 \times 10^{-4}\ \text{cm}^{-1} = 15{,}801.6\ \text{cm}^{-1}$. The sampling frequency to be used for the earth radiation spectral bandwidth is determined by the maximum wave number from the following equation:

$$\nu_s \geqq 2\nu_{max} \tag{5.24}$$

which is known as the Nyquist criterion. If the sensor is designed to measure radiation between 400 and 3800 cm^{-1}, we have

$$\nu_{max} = 3800 \quad \text{cm}^{-1} \tag{5.25}$$

so that

$$\nu_s = 7600 \quad \text{cm}^{-1} \tag{5.26}$$

The sampling interval is set to take interferogram data every four fringes, so $\nu_s = 15{,}801/2 = 7900$.

The white-light interferometer is designed to trigger the laser interferometer fringe-counting system to start counting by its zero optical path signal.

Some sensor designers change the interferometer spectrometer plane mirrors for the following optical systems: (a) corner cube reflector, (b) cat's-eye reflector, and (c) prism reflector. For space applications, cat's-eye and corner cube reflector systems offer one of the best solutions for vibration-free scanning of the moving mirror.

The interferometer spectrometer signal-to-noise ratio is calculated as shown in the following example:

$$S/N = LD^*\theta\sqrt{T\tau}/L_0\sqrt{A_D} \tag{5.27}$$

where L is the earth emission radiance, in watts per square centimeter reciprocal centimeter steradian, D^* the detector detectivity, in centimeters hertz$^{1/2}$ per watt, θ the system throughput, in square centimeters steradian, T the dwell time, in seconds, τ the system efficiency, L_0 the total optical path difference, in centimeters, and A_D the detector area, in square centimeters.

The throughput of the interferometer spectrometer is

$$\theta_I = \frac{2\pi A\ \Delta\nu}{\nu_{max}} \quad \text{cm}^2\ \text{sr} \tag{5.28}$$

where A is the area of the interferometer mirror, $\Delta\nu$ the spectral resolution, and ν_{max} the maximum wave number to be measured.

The throughput of the detector system is

$$\theta_D = A\Omega_D \quad \text{cm}^2\ \text{sr} \tag{5.29}$$

where the detector solid angle is 1.5 sr for an $f/1$ detector radiation collector system.

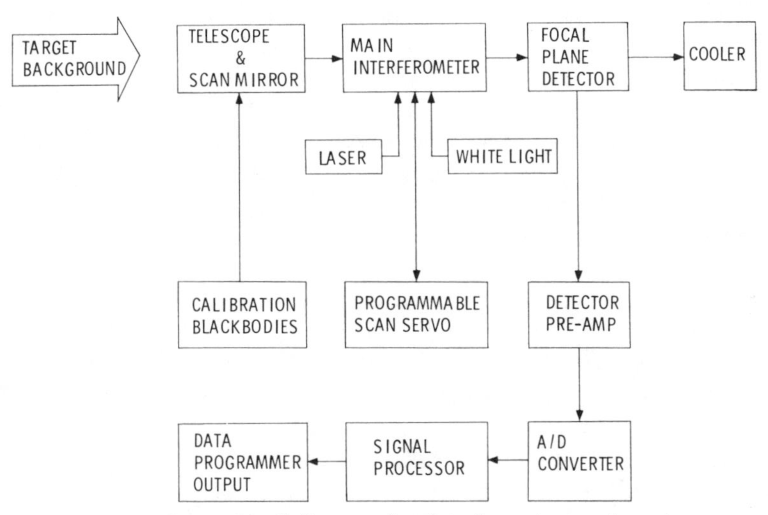

Fig. 5.8. System block diagram of an interferometer spectrometer.

Let the interferometer maximum wave number be 4000 cm^{-1}, the total optical path difference 10 cm, the interferometer aperture 2 in., the detector area 0.04 cm^2, and the detector solid angle 1.5 sr; the interferometer solid angle is calculated as 0.03 cm^2 sr, the detector detectivity is 10^{10} cm $Hz^{1/2}/W$, the system efficiency is 0.1, the dwell time is 1 sec, and the earth radiance is 300 K blackbody radiation at 1000-cm^{-1} wave number.

The signal-to-noise ratio at 0.1-cm^{-1} spectral resolution is calculated as

$$\begin{aligned} S/N &= (1)(10^{-5})(1)(10^{10})(0.03)(1)(0.1)/(10)(0.2) \\ &= 150 \end{aligned} \tag{5.30}$$

Figure 5.8 shows a block diagram of the Michelson interferometer spectrometer system.

5.6 Prism and Grating Materials

The best prism material for the prism spectrometer should have a large dispersion per wavelength, small refractive index change per temperature variation, and wide spectral range. In the visible region, there is a wide variety of optical glass available for the prism material. Uviol crown, silicate crown, light flint, and dense flint are good candidates for visible prisms. Liquid prisms have been used to some extent in the ultraviolet and visible region. They can be designed in large sizes and have uniformity, homogeneity, and a wide range of dispersions. The disadvantage of liquids is their high temperature coefficient of dispersion, so that more temperature control is required than for a solid prism. The liquids that have been most used in liquid prisms are water, and monobromonaphthalene. Water has great transparency in the ultraviolet near 2000 Å.

Many optical crystals have been used for prism materials, especially for application in the ultraviolet and infrared regions, where the transmittance of optical glass is not good. Quartz is one of the best crystals of large size and optical quality. It is the most widely used material, after glass, in UV, visible, and near-IR prism spectrometers. Table 5.8 gives the most useful prism materials for different spectral regions.

The most useful grating materials are fused silica and aluminum. The aluminum grating has higher reflectivity in the UV region. Since a solid piece of aluminum cannot be polished easily to an optical surface, the grating blank glass is coated by evaporation of chromium, then coated with an aluminum layer. The grating grooves are ruled into this aluminum

Table 5.8

Most Useful Prism Materials

Material	Spectral region (μm)
Quartz, SiO_2	0.4–2.0
Lithium fluoride, LiF	2.7–5.5
Fluorite, CaF_2	1.0–4.0
Rock salt, NaCl	8.0–16.0
Potassium bromide, KBr	3.0–20.0

layer for diffraction applications. Another way to obtain high grating reflectance is by evaporating aluminum onto a grating-ruled metal or glass. This method is very satisfactory even for a larger grating with 30,000 lines/in. and 8 in. of ruled surface. Gratings made for use in the UV and near infrared have sometimes been coated with gold to ensure longer life of the reflecting surface. Good replica grating are available that are copies of metal or glass gratings. They are made as collodion casts of the grating surfaces, mounted, after removal, on a silvered or transparent base of glass, which may be flat or concave. Developments in plastics have made possible the production of replica gratings equal in efficiency and resolving power to the master original from which the cast was made. For a large grating the replica has the advantage of low cost.

Gratings used in high-energy laser applications, especially with high-power space laser systems, must be good thermal conductors. This can be accomplished by replicating onto solid copper rather than glass substrates. Sometimes even this will not work, as with very high power lasers; the only solution then is the use of special gratings with grooves cut directly into a metallic blank.

Holographic gratings are produced by permitting two beams of coherent laser light to form interference fringes in a photosensitive material deposited on an optical substrate. The interference fringes are then developed and coated with an appropriate reflecting coating. This process can be used for the production of both plane and concave gratings. A new blazing technique has been developed in which grooves are ion-etched into the substrate material itself. This may be of great interest because a master ion-etched grating is highly damage-resistant and therefore well suited for use with high-energy space laser systems.

5.7 Interferometer Beamsplitter Materials

Every Michelson interferometer spectrometer requires some form of beamsplitter for radiation interference. The partially reflecting beamsplitter transmits and reflects half of the incident earth radiation to form the interferogram. The reflectance of a single plate can be improved by coating the reflecting surface with either a metal or a high-index dielectric material. An interferometer with a beamsplitter consisting of a coating on a supporting plate is unsymmetrical unless it is balanced by a compensating plate in the other split beam. When this plate does not match the support of the beamsplitter in both material and thickness, the two aberrations, astigmatism and chromatic aberration, are not compensated. The compensating plate corrects the major asymmetry caused by the support of the beamsplitter. Figure 5.9 shows the design of three different sym-

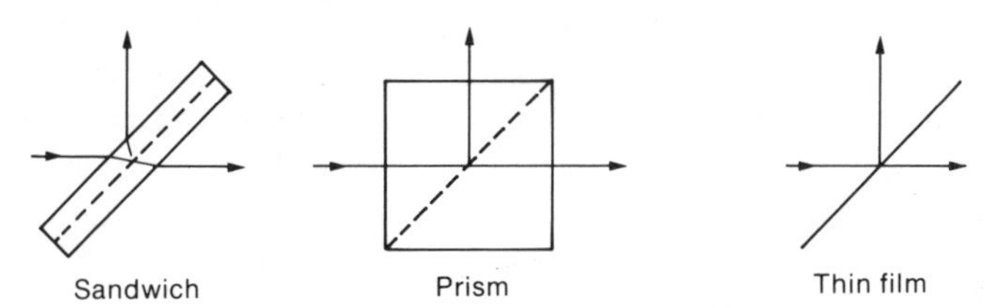

Fig. 5.9. Interferometer beamsplitter designs.

Table 5.9

Interferometer Beamsplitter Materials

Material	Spectral range (μm)
Quartz	0.4–2.0
CaF_2	1.0–4.0
ZnSe	0.6–15.0
KBr	3.0–20.0
CsI	10.0–50.0
Mylar	25.0–1000.0

metrical beamsplitters. Table 5.9 lists the wavelength ranges covered by various beamsplitter materials.

5.8 Space Spectrometer Sensors and Applications

5.8.1 *Space Prism Spectrometer Sensor*

The NASA Shuttle imaging spectrometer (SIS) is intended to provide opportunities for research and for advanced sensor concepts aboard the reuseable space transportation system (STS). The key to the design of the SIS is a sensor using two separate prism dispersing systems rather than a spectral filter to select the spectral bands. Earth radiation reflected from the surface is imaged on a slit, which defines the sensor field stop aperture and also determines the surface size on the ground. Since the prism spectrometer is designed to maintain the spatial resolution present at the entrance slit, the image at the spectrometer focal plane is a series of lines of the earth surface images that can be measured by CCD array detectors. Within a measurement time interval, the STS orbital motion moves the line image into the next line position so that the imaging process can be repeated in a pushbroom-type scanning fashion.

The optical design approach is that the reflected earth radiation enters the sensor through an 11-cm-aperture reflecting Schmidt corrector plate, which is located near the center of curvature of the primary mirror. Radiation reflected from the Schmidt corrector illuminates the primary mirror to produce an image at the primary focal plane, where a reflective slit defines the field stop and serves as the entrance slit for the prism spectrometer. The reflective slit illuminates the primary mirror, which now serves as the collimator. The radiation then passes to the dispersing section, which is comprised of a beamsplitter, prisms, and the Schmidt corrector mirror for the spectrometer system. The radiation is refocused through the system again by the primary mirror at the final focal plane of the sensor. The spectrometer system focal plane consists of two separate focal plane subassemblies: a silicon CCD for visible and near-infrared radiation measurement and an HgCdTe CCD for short-wavelength infrared radiation measurements. The SIS sensor parameters are summarized in Table 5.10.

Table 5.10

SIS Sensor Parameters

Orbit height	250 km
Swath width	12.1 km
RFOV	30 m
TFOV	2.8°
Spectral bands	0.4–2.5 μm, 128 bands
Spectral bandwidth	10 nm (VNIR), 20 nm (SWIR)
Telescope aperture	11 cm, Schmidt type
f-number	*f*/3.8
Focal plane detector	
VIS/NIR	Silicon CCD, 64 × 404 pixels
SWIR	HgCdTe CCD, 6–64 × 64 chips
Detector temperature	120 K
Date rate	105 Mbps

5.8.2. *Space Grating Spectrometer Sensors and Applications*

5.8.2.1 *The Satellite Infrared Spectrometer*

The Satellite Infrared Spectrometer (SIRS) has been developed for the purpose of remote sounding of vertical atmospheric temperature and water vapor profiles. In its ultimate application, the sounding results from this sensor would be used to determine the temperature and water vapor fields of the earth's atmosphere.

The SIRS sensor is basically a multispectral, *f*/5 Ebert–Fastie grating infrared spectrometer. The sensor FOV is alternately switched between the earth scene and space at 15 Hz, using a two-blade reflective chopper. The diffracted energy from the grating is reflected by the primary mirror and is focused at the exit slit focal plane. Multiple exit slits, order filters, and detectors are placed at positions in the exit slit focal plane to accept energy at the desired wavelength interval. The diffraction grating is used in the first four orders to accommodate the large wave number spread encompassed by the required 14 spectral bands. The equivalent radiation bandpass for these bands is equal to the desired 5 cm^{-1}.

In a grating spectrometer, it is necessary to filter out undesired orders. For SIRS, this is accomplished with order filters consisting of a combination of interference and bulk absorption filters.

Table 5.11 lists the sensor system parameters. Figure 5.10 shows the SIRS sensor schematic, and Fig. 5.11 shows temperature sounding results from SIRS space measurements.

Table 5.11

SIRS Sensor System Parameters

Sensor type	Ebert–Fastie grating spectrometer	
Focal length	31.75 cm	
f-number	5.0	
IFOV	0.04 sr	
Spectral bands	899.0 cm^{-1}	order 4
	750.0	3
	714.0	3
	706.0	3
	699.0	3
	692.0	3
	679.8	3
	668.7	3
	531.0	2
	436.0	2
	436.5	2
	425.5	2
	302.0	1
	291.5	1
	280.0	1
Spectral resolution	5.0 cm^{-1}	
Grating ruling	26.93 lines/mm	
Detectors	Immersed thermistor bolometer	
Dynamic range	800	
Weight	70 lb	
Power	30.6 W	

5.8.2.2 *The Solar Backscatter Ultraviolet and Total Ozone Mapping Spectrometers*

The solar backscatter ultraviolet (SBUV) and total ozone mapping spectrometers are composed of grating spectrometers. The SBUV sensor system consists of a double Ebert–Fastie spectrometer. This spectrometer monitors 12 selected narrow wavelength bands in the spectral region from 250 to 340 nm, or continuously scans the wavelength range from 160 to 400 nm to determine total ozone and its vertical concentration distribution above the ozone maximum by measuring the scattered solar radiance.

The spectrometer has a spherical collimator mirror of 25-cm focal length with an aluminum and holographic diffraction grating of 5.2 × 5.2 cm area and 24,000 grooves/cm. The spectrometer has fixed entrance and exit slits that are 3.0 cm long. Their widths are equivalent to a 1-nm spectral bandwidth near 0.3 μm. It can scan continuously the wavelength range from 160 to 400 nm in 0.2-nm steps. A depolarizer is used to elimi-

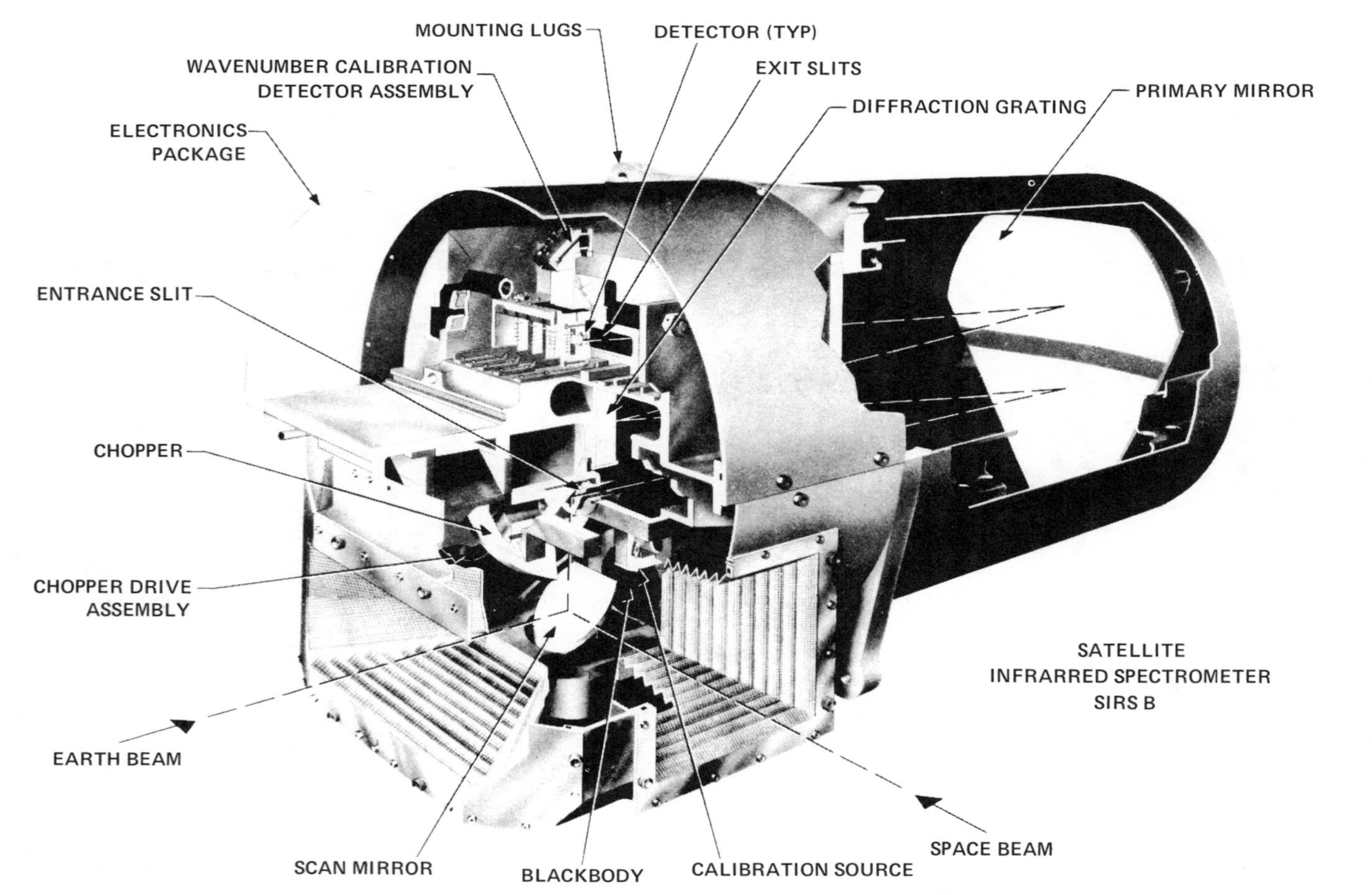

Fig. 5.10. The Nimbus satellite infrared spectrometer (SIRS). (Courtesy of Santa Barbara Research Center.)

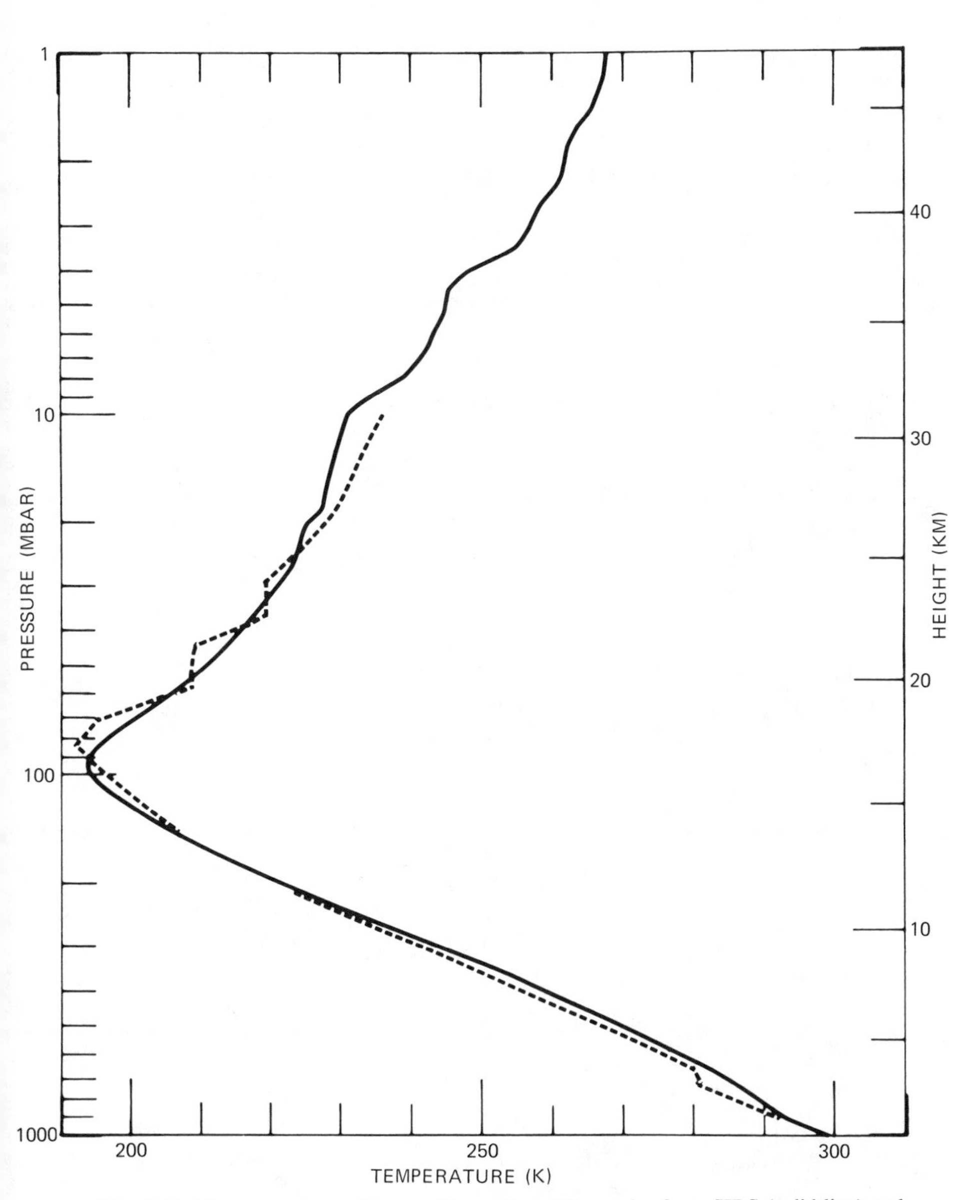

Fig. 5.11. Temperature profile over Green Bay, Wisconsin, from SIRS (solid line) and radiosonde (dashed line). (Courtesy of D. Q. Wark and NASA.)

nate the sensitivity of the grating spectrometer to polarization of the backscattered UV solar radiation.

Two types of detectors are used for the sensor: a photodiode and a photomultiplier tube (PMT). The PMT is an ITT F4090 type with a bialkali photocathode and a fused silica window. The window material was selected for transmittance at 160 nm and minimum fluorescence. The FOV for the SBUV is 11.3 × 11.3°, which traces 200-km-wide swaths on the ground. Figure 5.12 shows an optical diagram of the SBUV. Figure 5.13 shows the ozone number density measured by the SBUV from space.

The total ozone mapping spectrometer (TOMS) is also an Ebert–Fastie spectrometer with a fixed grating and an array of exit slits. The spectrometer has an FOV of 3 × 3° and measures six bands from 321.5 to 380 nm with a spectral bandwidth of 1 nm. A scanning mirror scans across the track 51° from the nadir in 3° steps for both sides. The total time at each scene during scanning and stepping is 200 msec, which includes 168 msec for data sampling (four samples each in six bands) and 32 msec for the scanning mirror setting. Figure 5.14 is an optics diagram of TOMS.

5.8.2.3 *The Coastal Zone Color Scanner*

The coastal zone color scanner (CZCS) on Nimbus 7 is the first space sensor devoted to the measurement of ocean color. Its objectives are to determine the contents of water as quantitatively as possible, to carry out

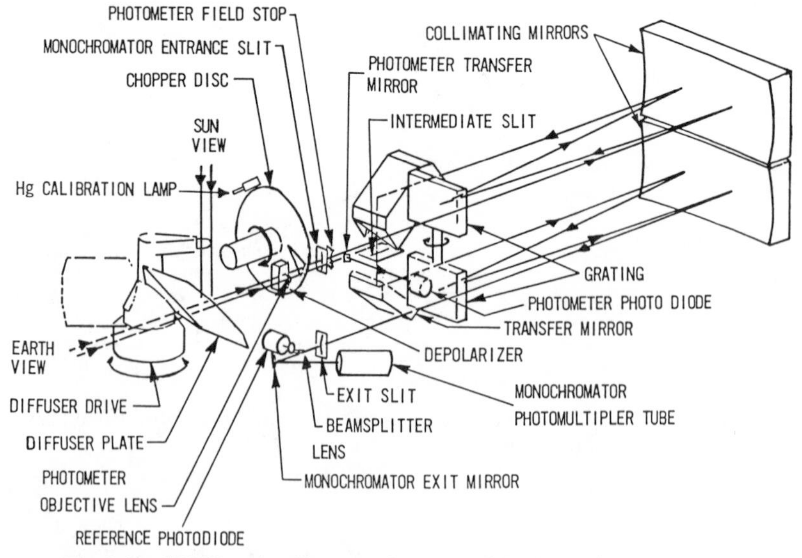

Fig. 5.12. SBUV optics diagram. (Courtesy of Beckman Co. and NASA.)

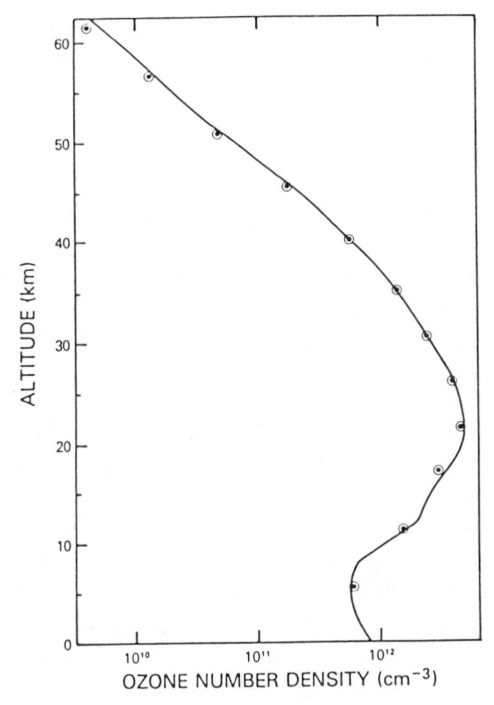

Fig. 5.13. Ozone number density measured by the SBUV (circled dots). This is compared with the calculations of Krueger–Minzner (solid lines). (Courtesy of R. D. McPeters, NASA.)

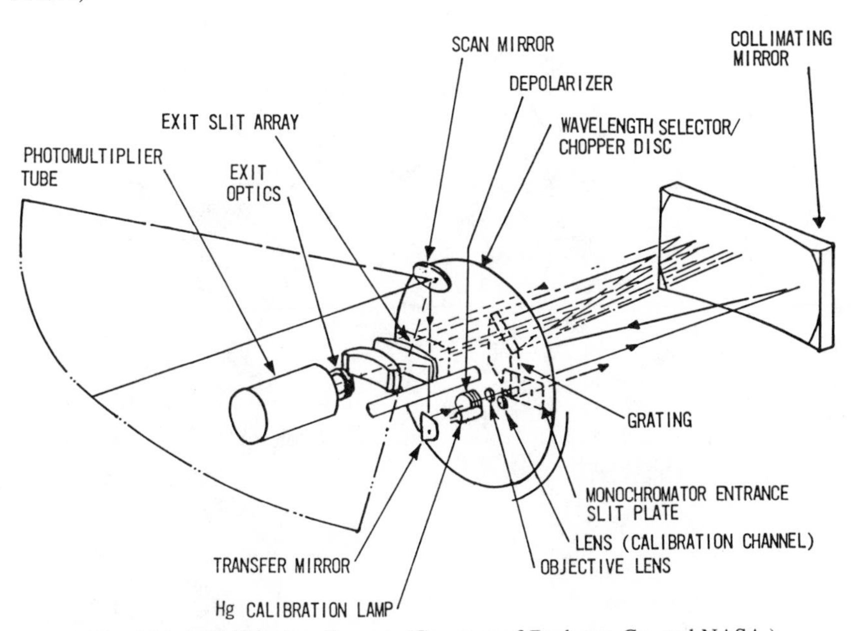

Fig. 5.14. TOMS optics diagram. (Courtesy of Beckman Co. and NASA.)

such measurements over a global area, and to discriminate between organic and inorganic materials in the water. The CZCS data processing goal is to take the observed radiance, determine the radiance that would be sensed directly above the ocean surface, and then derive from the measured radiance the content of the water below the ocean surface.

The CZCS is a scanning multispectral space sensor with a swath width of 1566 km centered on spacecraft nadir. The RFOV is 0.825 km at nadir. The sensor has six spectral bands, five sensing solar radiation backscattered from the ocean surface and one sensing thermal radiation emitted from the ocean. The incident radiation is split with a dichroic beamsplitter; one beam is dispersed by a grating and detected by five silicon diode detectors at the exit slit of the spectrometer system. The ocean surface radiation in the 10.5–12.5-μm spectral band is reflected from the beamsplitter and imaged onto an infrared HgCdTe detector at 120 K. The first four channels were selected to cover specific absorption bands of the ocean surface. These bands are designed to look at water only and saturate when looking over most land surfaces and clouds. Figure 5.15 shows the sensor system arrangement. Table 5.12 shows the center wavelengths, spectral bandwidths, and minimum signal-to-noise ratio for the sensor at the darkest surface radiances.

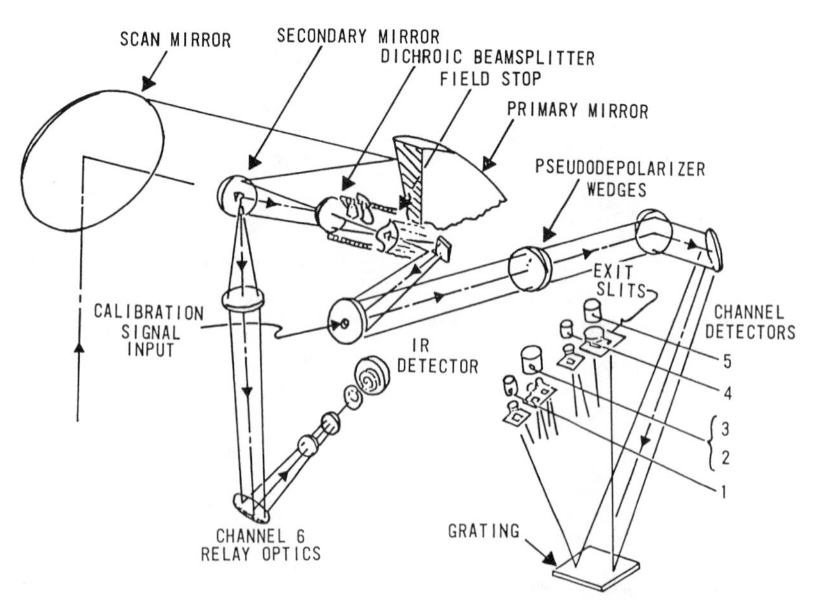

Fig. 5.15. CZCS optical arrangement. (Courtesy of Ball Bros. Co. and NASA.)

Table 5.12

CZCS Performance Parameters[a]

Performance parameters	Channels					
	1	2	3	4	5	6
Scientific observation	Chlorophyll absorption	Chlorophyll correlation	Yellow stuff	Chlorophyll absorption	Surface vegetation	Surface temperature
Center wavelength λ, μm	0.443 (blue)	0.520 (green)	0.550 (yellow)	0.670 (red)	0.750 (far red)	11.5 (infrared)
Spectral bandwidth $\Delta\lambda$, μm	0.433–0.453	0.510–0.530	0.540–0.560	0.660–0.680	0.700–0.800	10.5–12.5
Instantaneous field of view (IFOV)	0.865 × 0.865 mrad (0.825 × 0.825 km at sea level)					
Coregistration at nadir	<0.15 mrad					
Accuracy of viewing position information at nadir	<2.0 mrad					
Signal-to-noise ratio (min.) at radiance input $N <$ (mW/cm^2 sr μm)	>150 at 5.41	>140 at 3.50	>125 at 2.86	>100 at 1.34	>100 at 10.8	NETD of 0.220 K at 270 K
Consecutive scan overlap	25%					
Modulation transfer function (MTF)	1 at 150-km target size, 0.35 min. at 0.825-km target size					

[a] Courtesy of NASA.

5.8.2.4 *Spacelab-1 Grille Spectrometer*

A grille spectrometer was launched as part of the payload of Spacelab-1. The sensor was used to measure atmospheric species in the infrared region from 2.5 to 13μm with a spectral resolution better than 0.1 cm^{-1}. Most of the observations were performed in the solar viewing earth atmospheric absorption mode with fast spectral scanning during the sunset and sunrise periods. Some of the measurements were made in the earth thermal emission mode by observing the earth's atmospheric limb at almost any time during the night or day. The radiation coming from the sun through the earth's atmospheric limb or from the atmospheric limb itself is reflected toward a telescope by an orientable rectangular plane mirror. The telescope, which transmits the radiation to the spectrometer, has a 0.3-diameter and 6-m focal length. Two detectors are used simultaneously to cover the entire 2.5–13-μm range. All functions of the sensor are programmable through a microprocessor, which is part of the sensor electronics and allows interaction between the payload specialist on board the Shuttle and ground-based scientists. The spectrometer measures vertical concentration profiles of atmospheric species from 15 to 150 km in altitude.

5.8.2.5 *Stratospheric Aerosol and Gas Experiment Spectrometer*

The stratospheric aerosol and gas experiment (SAGE I) sensor is a grating spectrometer type that measures the extinction of solar radiation during solar occultation in four bands centered at 0.385, 0.45, 0.6, and 1.0 μm. Radiançe measurements from the sensor can be used to determine vertical profiles of aerosol extinction as well as concentration profiles of ozone and other gases. The sensor consists of an optical and an electronics module. The optical module consists of a flat scanning mirror, telescope, diffraction grating, and four filtered silicon detectors. The SAGE sensor measures the attenuation of solar radiation by the limb of the earth's atmosphere in the four bands during each spacecraft sunrise and sunset (15 cycles in each 24 hr). The detector output signal is converted to an attenuation coefficient versus vertical height. Each sunrise and sunset are monitored by the SAGE sensor from the top of the clouds to 60-km altitude. Figure 5.16 shows a diagram of the SAGE I sensor optical system. The SAGE I sensor is the payload of atmosphere explorer mission B (AEM-B).

The SAGE II space sensor is the second generation of SAGE flown on the AEM spacecraft. The sensor is designed to monitor global vertical profiles of stratospheric aerosols, ozone, and nitrogen dioxide by measuring the scattering and absorption of solar radiation through the earth's

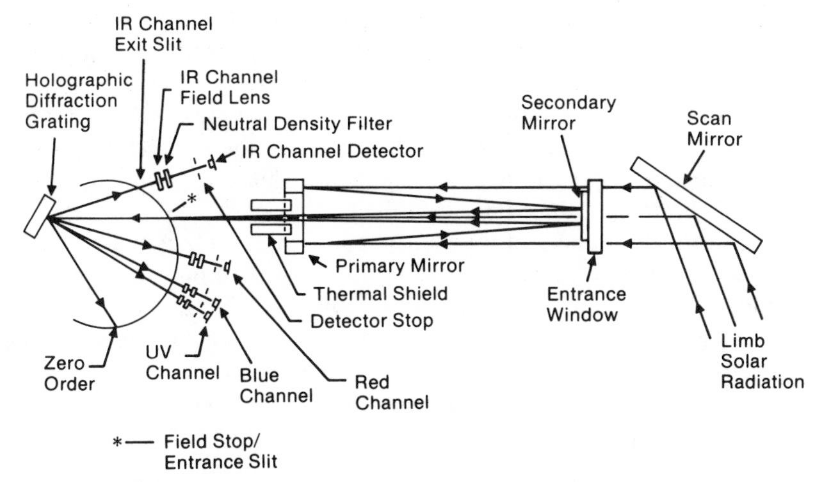

Fig. 5.16. SAGE sensor optical schematic. (Courtesy of Ball Bros. Co. and NASA.)

atmosphere during spacecraft solar occultations for climatic and environmental applications.

During satellite sunrise and sunset events, an azimuth gimbal tracks the sun and the entire sensor optical system rotates with the azimuth gimbal. An elevation scan mirror provides vertical up-and-down scanning motion. Solar radiation reflected from the scan mirror enters a Cassegrain telescope, which forms an image on the entrance slit of a grating spectrometer. This slit defines the instantaneous field of view of the sensor. Scattered and absorbed solar radiation through this slit aperture is dispersed by a concave holographic grating, and the diffracted beams along the Rowland circle form seven spectral bands (0.385, 0.448, 0.453, 0.525, 0.60, 0.94, and 1.02 μm). All seven bands have silicon photodiode detectors used in the photovoltaic mode. Each band is taken typically from the cloud tops up to 150-km tangent height in the atmosphere. The SAGE II sensor is the payload of ERBS.

5.8.3 *Space Interferometer Spectrometer Sensors and Applications*

5.8.3.1 *The Fabry–Perot Interferometer Spectrometer*

5.8.3.1.1 *The High-Resolution Doppler Imager (HRDI).* This is a new triple-étalon Fabry–Perot interferometer that collects the visible radiation emitted from the earth's atmosphere to obtain the temperature from the emission line width and wind field from the line shifts from the upper troposphere through the thermosphere.

The HRDI sensor consists of an optical telescope and two-axis pointing system, a triple-étalon Fabry–Perot interferometer with support electronic systems. The telescope system is approximately 1.8 m in length and 20 cm in diameter. It is mounted on a two-axis pointing system viewing 45, 135, 225, and 315° in azimuth with respect to the spacecraft velocity vector. The telescope can also be scanned vertically to obtain vertical profiles of the temperature and wind field.

The central component of the instrument is the étalon, which performs the high-resolution spectral analysis by multiple path interference of the transmitted atmospheric radiation. Wavelength selection is performed to use an image plane detector, which in effect provides simultaneous angle scanning of the spectral intervals of the emission lines.

Atmospheric radiation enters the interferometer from a light pipe coupled to the telescope focal plane. A collimating lens forms the light pipe divergence angle to meet dielectric interference filter requirements. A dual filter wheel is used to provide enough filters and blocks to cover the many wavelengths required for the mission. After the dielectric filters, a collimating double-lens system is used to narrow the divergence angle down to the 0.4° half-angle required by the étalons. Radiation passes through the low- and then medium-resolution étalons. These two etalons have piezoelectric spacers to allow spectral selection by varying étalon spacing distance. The radiation then passes through the fixed-spacing high-resolution étalon (HRE) and is focused at the focal plane as a concentric ring spectral image to the image plane detector.

The detector concentric anode layout is designed to match the constant-area, concentric ring output of the Fabry–Perot. The detector has a fused silica faceplate of 32-mm effective diameter with an extended red multialkali S-20 photocathode and the 32 anodes that are used to measure the Fabry–Perot rings by the sensor system.

The HRDI sensor is one of the payloads of NASA's upper atmospheric research satellite program. Table 5.13 lists the HRDI sensor system parameters.

5.8.3.1.2 *Cryogenic Limb Array Étalon Spectrometer (CLAES).* This sensor is designed to obtain global measurements of the concentrations of a set of upper atmospheric species of prime interest for earth environmental and climate monitoring. Vertical temperature and species concentration in the 15–65-km range can be determined from measurements of limb radiance by the CLAES sensor.

The sensor has an off-axis Gregorian telescope that uses knife-edge baffles, low-scattering superpolish mirrors, and Lyot stop for better opti-

Table 5.13

HRDI Sensor System Parameters

Telescope aperture	17.8 cm
IFOV	0.1° × 1.6° or 0.1 × 0.5°
Looking angle	Az: 45°, 135°, 225°, 315°
	Ze: 105°–115°, 116°–150°
Spectral bands	0.5577 μm
	0.6300 μm
	0.6875 μm
	0.7320 μm
	0.7635 μm
Étalons	High-resolution, 0.198 Å
	Medium-resolution, 1.10 Å
	Low-resolution, 6.0 Å
Detector	Special electron multiplier
Weight	76 kg
Power	88 W
Date rate	4.7 kbps

cal performance. Each atmospheric species is measured by means of a unique combination of angular scanning solid étalons and fixed blocking filters. The sensor concept is illustrated in Fig. 5.17. Atmospheric radiation measurement depends on the selection of any one of four angular scanning, cooled solid étalons, in combination with one or more of eight blocking filters. The étalon wheel approach was adopted because a mechanical rotation paddle action can be used both to select a given étalon and to scan it when it is in position. The spectrally filtered atmospheric radiation is focused on a cryogenically cooled focal plane consisting of a set of infrared detectors.

A Fabry–Perot étalon interference spectrometer was chosen for CLAES because it provides the largest throughput and high spectral resolution for the sensor system. There is a single-stage solid-hydrogen cooler in which the spectrometer unit and detectors are conductively cooled by direct contact with the cryogen tank, and the telescope system is vapor-cooled to 130°K.

The sensor views the earth's limb from orbit, sensing in a direction perpendicular to the spacecraft velocity vector and away from the sun vector. Altitude coverage and vertical spatial resolution are provided by the linear array of 20 detectors. Each of the main array detectors projects a 2.8-km vertical resolution at the limb; the full detector array covers

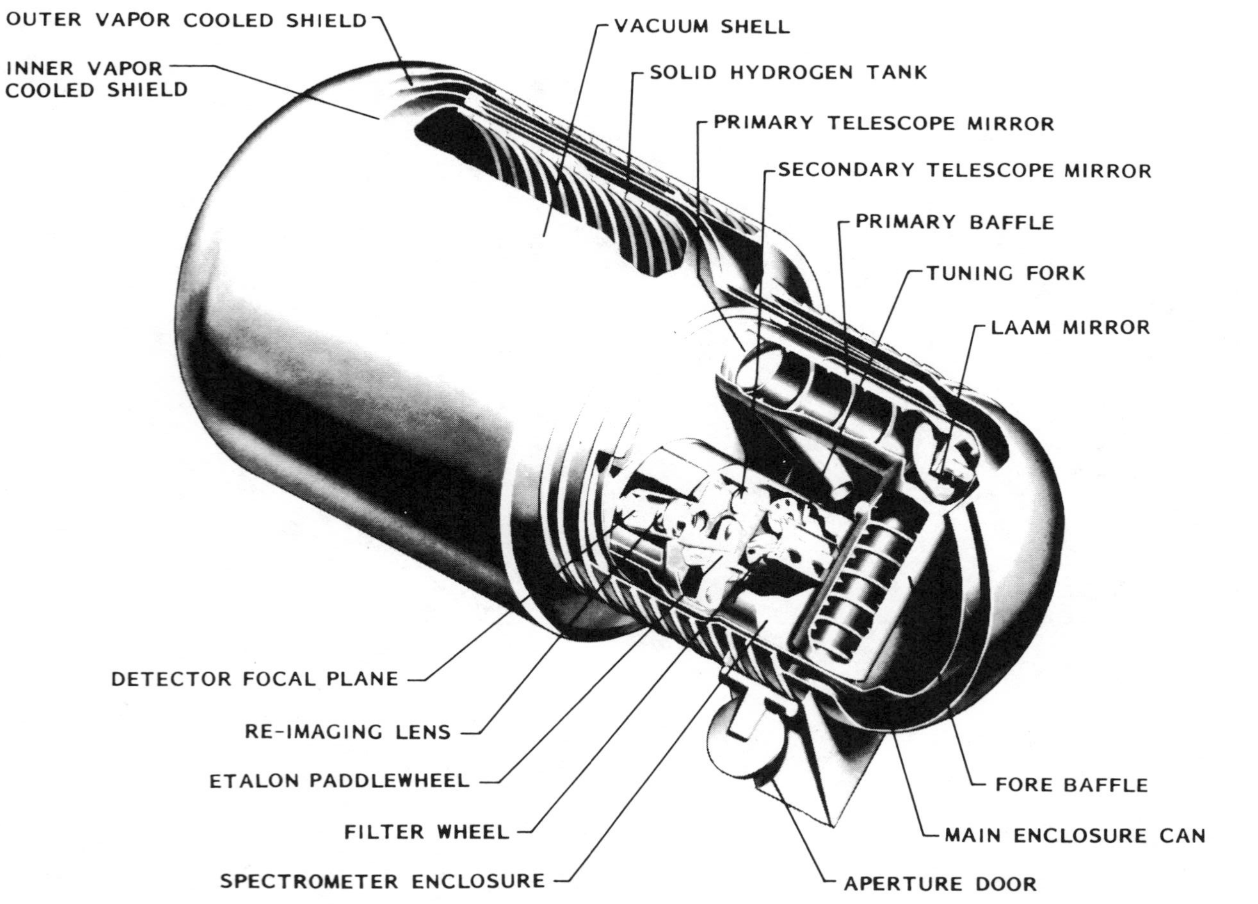

Fig. 5.17. CLAES sensor general arrangement. (Courtesy of Lockheed and NASA.)

approximately 50 km. The intent of the multidetector array is to avoid the need for a scanning mirror, thereby simplifying the optomechanical system and reducing demands on the cryogen. The horizontal instantaneous footprint of the main array detectors is 8.4 km, but the pushbroom horizontal resolution is a function of the length of time taken to complete a given étalon angular scan. The scan sampling horizontal length varies from 30 to 84 km for atmospheric species. Table 5.14 lists the CLEAS sensor parameters.

Table 5.14

CLAES Sensor Parameters

Telescope	
Aperture	15 cm, $f/3$
IFOV	1×3 mr
RFOV	2.8×8.4 km
Total vertical FOV	17.5 mrad
Off-axis rejection	10^{-6} at 0.6°
Spectrometer	
Aperture	5 cm
Spectral range	830–3300 cm^{-1}
Spectral resolution	0.25–0.5 cm^{-1}
Number of étalons	4
Spectral band filters	8
Detectors	
Type	Gallium-doped silicon
Number	
Main array	20
HCl array	3
Size	
Main	18×54 mil
HCl	90×54 mil
Cryogenics	
Type	Solid hydrogen
Volume	944 liters
Temperature	
Telescope	130 K
Baffle	130 K
Spectrometer	12 K
Detector	12 K
Lifetime	2 years
Envelope	1.23-m diameter, 2.57-m length
Weight	450 kg
Power	20 W
Data rate	3 kbps

5.8.3.2 *Michelson Interferometer Spectrometer*

5.8.3.2.1 *The Infrared Interoferometer Spectrometer (IRIS).* The IRIS sensor is designed to provide information on the earth's vertical temperature profile, water vapor profile, and ozone profile on a global scale to be used in climate and environmental studies. The sensor is a Michelson interferometer spectrometer type. The first part of the interferometer is the beamsplitter, which divides the incoming radiation into two nearly equal components. The sensor beamsplitter is made of potassium bromide, KBr. It has a multilayer dielectric coating that is optimized to the 5–20-μm region. The fringe-counting interferometer generates sine waves of 625 Hz at the silicon detector from a neon discharge lamp. The interferometer moving mirror has a drive coil and a pickup coil to generate a voltage proportional to mirror velocity. The velocity signal is also used in a feedback arrangement to provide electric damping and to make the system less sensitive to external vibration.

All mirrors of the interferometer are gold-coated. The fixed interferometer mirror is the entrance pupil for the sensor system. It has an effective circular aperture of 3.5-cm diameter. An ellipsoidal mirror collects the energy from the interferometer and focuses it onto the infrared detector, which serves as the exit pupil. An image motion compensation mirror is used in the sensor system to reduce the sounding area smear problem. Sounding area smear is minimized by slowly rotating the image motion compensation mirror while in the earth-viewing position through 3.5° in 10 sec, which is the scan time of an interferogram. Each interferogram represents a radiation field that originates within a well-defined area of about 145-km diameter of the atmosphere and the earth's surface.

The spectral resolution for IRIS-1 was 5 cm^{-1}, and the IRIS-2 moving mirror had to be scanned 0.36 cm for a better spectral resolution of 2.8 cm^{-1}. The larger moving mirror displacement of IRIS-2 required either more sample points per interferogram or a larger sampling separation between interferograms. The spectral range of IRIS-2 was reduced from 2000 to 1600 cm^{-1} to permit an adequate and comfortable oversampling margin.

The IFOV was reduced from 4° to 2.5° half cone angle for IRIS-2. The circular sounding area under measurement during the scanning of the interferogram was reduced from a diameter of 145 km to 95 km, which increased the probability of sensing homogeneous atmospheric and surface conditions. The spectral analysis is easier and more precise than for spectra representing mixed clear atmosphere and clouds of different kinds at different altitudes.

Table 5.15 lists the IRIS-1 and IRIS-2 sensor parameters. Figure 5.18

Table 5.15

IRIS-1 and IRIS-2 Sensor Parameters

Parameters	IRIS-1	IRIS-2
Spacecraft	Nimbus 3	Nimbus 4
Spectral range, cm^{-1}	400–2000	400–1600
Spectral resolution, apodized, cm^{-1}	5	2.8
Sensor aperture, cm	13	15
IFOV, deg	8	5
Interferogram scan time, sec	10.956	13.107
Moving mirror displacement, cm	0.2	0.36
Total optical path, cm	0.4	0.72
Velocity of scan mirror, cm/sec	0.0184	0.0275
Number of samples per interferogram	3408	4096
Frequencies in data channel, Hz	13–73	11–88
Detector	Therm/bolometer	Therm/bolometer
D^*, cm $Hz^{1/2}$/W	10^9	10^9
Detector area, cm^2	3.0 × 3.0	2.3 × 2.3
Operation temperature, K	250	250
Weight, kg	14	23
Power, W	16 (average)	28 (average)

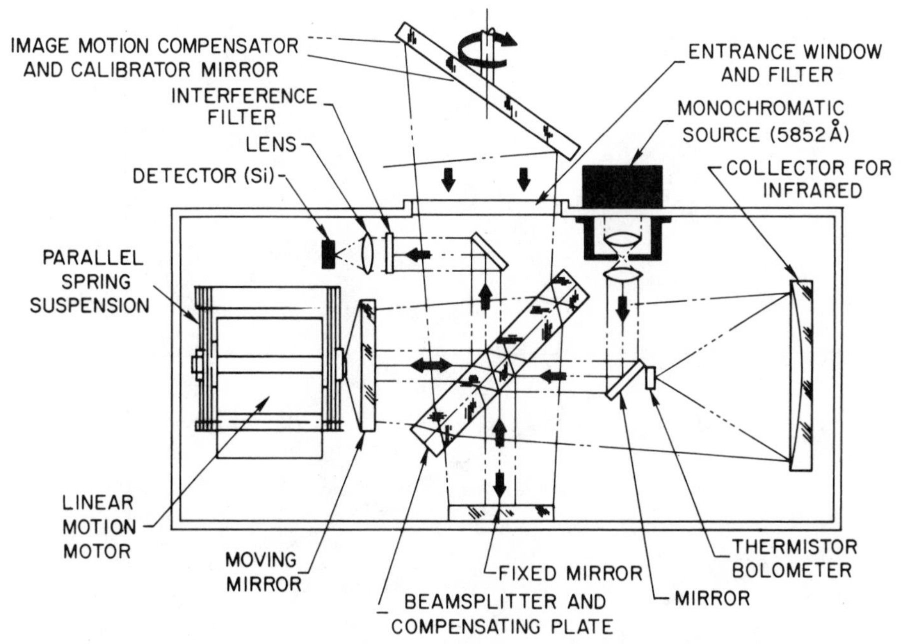

Fig. 5.18. Schematic of Nimbus infrared interferometer spectrometer. (Courtesy of Texas Instruments and NASA.)

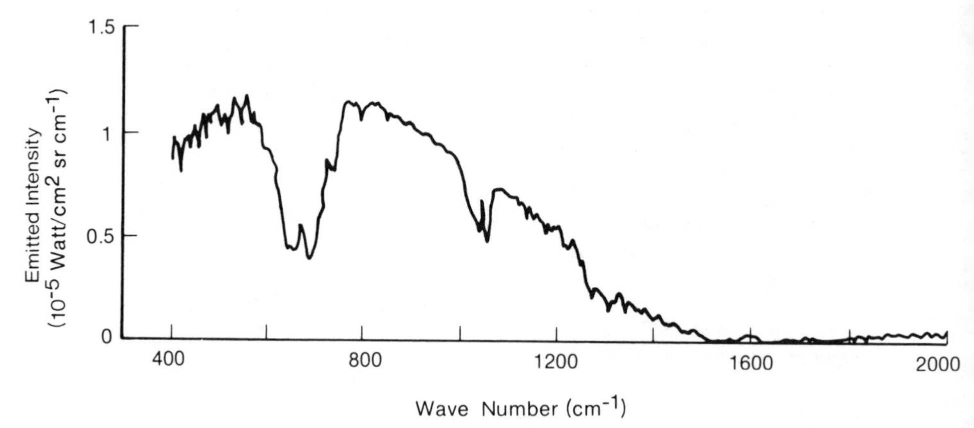

Fig. 5.19. Thermal emission spectrum of the earth measured by IRIS. (Courtesy of R. A. Hanel and NASA.)

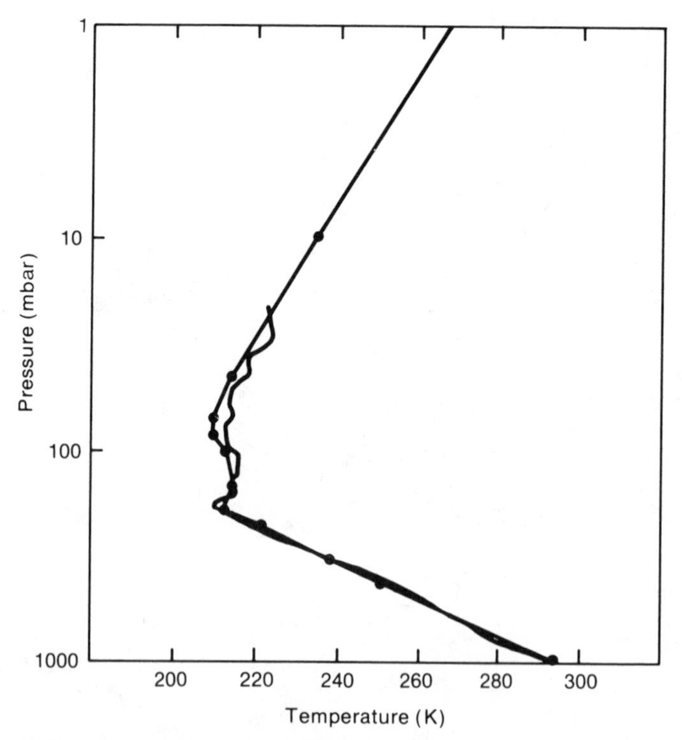

Fig. 5.20. Vertical temperature inversion from IRIS spectrum. IRIS, dots; Radiosonde, solid curve. (Courtesy of R. A. Hanel and NASA.)

shows a schematic diagram of the IRIS interferometer. Figure 5.19 shows a spectrum measured from space; Fig. 5.20 shows the temperature inversion result from the spectrum, and Fig. 5.21 shows water vapor inversion from the same spectrum.

5.8.3.2.2 *The Atmospheric Trace Molecular Spectroscopy Sensor (ATMOS).* This sensor is a multiflight Shuttle pallet Spacelab payload designed to determine the spatial variations of species in the upper atmosphere by the solar occultation technique. The sensor views the sun through the earth's atmosphere during Shuttle sunrise and sunset in order to produce high-resolution infrared absorption spectra.

The ATMOS sensor receives incoming radiation from the sun via a two-axis servo-controlled suntracker that automatically corrects for orientation changes in the relative position of the sun to an accuracy of 0.4 mrad with a stability of 0.06 mrad. The solar radiation passes through a ZnSe entrance window to the main sensor of ATMOS, which is a cat's-eye interferometer spectrometer. The beamsplitter substrate is KBr with a Ge/KRS-5 coating designed for the spectral region 2–16 μm. For absorbed solar radiation entering the interferometer, the optical path is var-

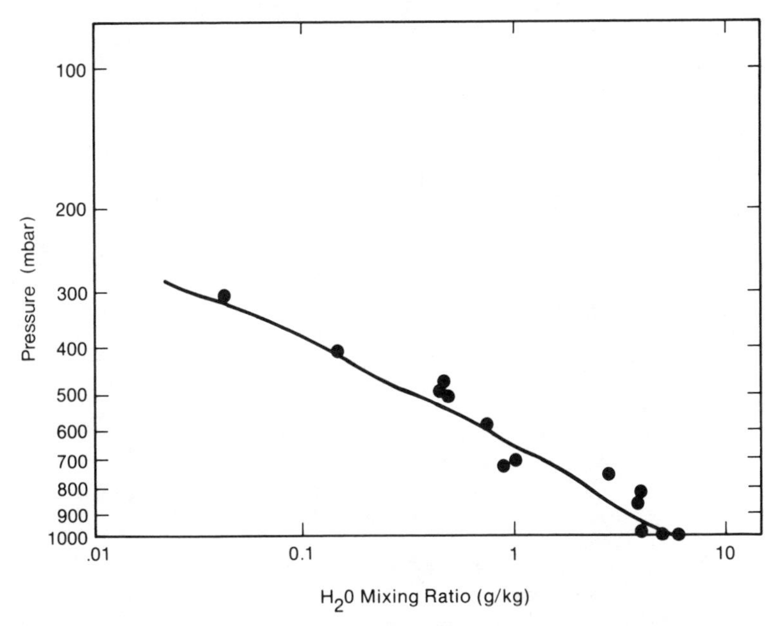

Fig. 5.21. Vertical water vapor inversion from IRIS spectrum. Radiosonde, dots; IRIS, solid curve. (Courtesy of R. A. Hanel and NASA.)

Table 5.16

ATMOS Sensor Parameters

Telescope aperture	7.5 cm, *f*/3
Spectral range	2–16 μm
Spectral resolution	0.01 cm^{-1}
Total optical difference	100 cm
IFOV	1.2 or 4 mrad
RFOV	2 km
Scan time	1 sec, single-sided 2 sec, double-sided
Scan direction	Bidirectional
Signal frequencies	31.2–250 Hz
Sampling frequencies	263,000–395,000 Hz
Detector	HgCdTe
Detector temperature	77 K
Data rate	16 Mbps
Weight	250 kg
Power	195 W (average)

ied continuously from −50 to +50 cm at a speed of 50 cm/sec by moving both cat's-eye optical systems simultaneously at equal speed but in opposite directions to produce the double-sided interferogram.

The laser interferometer used in ATMOS is a single-mode, single-frequency stabilized HeNe laser for sampling and spacing counting applica-

Table 5.17

HIS Sensor Parameters

Telescope aperture	40.6 cm, *f*/6
IFOV	0.28 mrad
RFOV	10 km
Spectral band	
Band 1	620–750 cm^{-1}
Band 2	890–1510 cm^{-1}
Band 3	2180–2700 cm^{-1}
Spectral resolution	0.36, 2.1 cm^{-1}
Optical path difference	1.4 cm (maximum)
Band 1	0.0–0.24, 0.59–0.70, 1.26–1.4 cm
Band 2	0.0–0.24 cm
Band 3	0.0–0.24 cm
Dwell time	0.48, 0.88, 4.48 sec
Scan mirror aperture	2.7 cm
Beamsplitter material	ZnSe
Detector	HgCdTe
Detector temperature	90 K

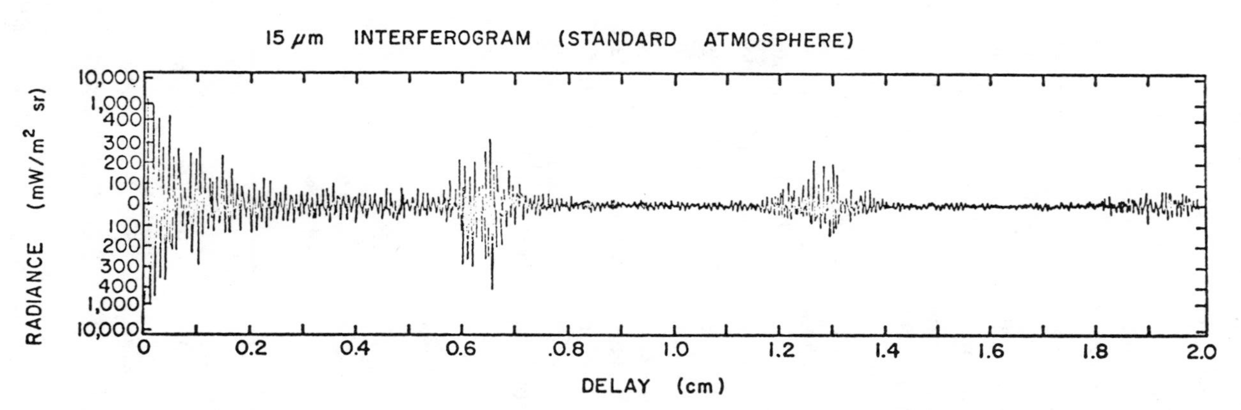

Fig. 5.22. Interferogram for the 15-um CO_2 band (600–750 cm^{-1}) computed for the U.S. standard atmosphere. (After Smith *et al.*, 1979.)

tion. The laser is sealed in a housing at 1-atm pressure for long-life operation in flight. Table 5.16 lists the parameters of the ATMOS sensor.

5.8.3.2.3 *The High-Resolution Infrared Sounder (HIS).* This sensor is designed as a GEO sounding instrument capable of achieving 1°C temperature accuracy, 10% moisture accuracy, and the highest vertical resolution achievable with passive space sensor techniques. The interferogram for the 15-μm atmospheric CO_2 radiation is not spread out, but instead is clustered in several relatively narrow optical path difference intervals due to the uniform spacing of the CO_2 absorption lines. These regions of peak signal amplitude occur at intervals of the optical path difference that are multiples of 0.6 cm, the reciprocal of the wave number spacing of CO_2 absorption lines. Thus, for the 15-μm CO_2 band it is not necessary to measure the complete interferogram to gather all spectral components of the upwelling radiation due CO_2 emission. It is only necessary to sample the interferogram in the three relatively narrow optical path difference intervals where the signal amplitude is significant. By scanning the moving mirror of the interferometer only in the optical path difference regions that contain the desired information, a significant savings in useful dwell time can be obtained.

The HIS sensor is designed with a path difference drive system that can scan rapidly only in the desired regions. Table 5.17 lists the sensor parameters. Figure 5.22 illustrates the computed interferogram for the 15-μm CO_2 band for temperature sounding applications.

General References and Bibliography

Ackerman, M., and Girard, A. (1978). Grille spectrometer. Spacelab Mission 1 Experiment Descriptions. *NASA Tech. Memo.* **NASA TM-82448,** 15–17.

Bell, R. J. (1972). "Introductory Fourier Transfer Spectroscopy." Academic Press, New York.

Girard, A. (1963). Spectrometre a grille. *Appl. Opt.* **1,** 79–87.

Hanel, R. A. (1969). The infrared interferometer spectrometer (IRIS) experiment. *In* "Nimbus 3 User's Guide," pp. 109–146. Goddard Space Flight Cent., Greenbelt, Maryland.

Hanel, R. A., and Conrath, B. J. (1969). Interferometer experiment on Nimbus 3: preliminary results. *Science* (*Washington, D.C.*) **165,** 1258–1260.

Hanel, R., Conrath, B., and Schachman, B. (1970). The infrared interferometer spectrometer (IRIS) experiment. "Nimbus 4 User's Guide." Goddard Space Flight Cent., Greenbelt, Maryland.

Hays, P. B. (1981). Instrument requirements document for the high resolution Doppler imager. *In* "Upper Atmosphere Research Satellites." Goddard Space Flight Cent., Greenbelt, Maryland.

Hays, P. B., Killeen, T. C., Kennedy, B. C. (1981). The Fabry–Perot Interferometer on Dynamics Explorer, *Space Sci. Instrum.* **5,** 395–416.

Heath, D., and Krueger, A. (1978). The solar backscatter ultraviolet (SBUV) and total ozone mapping spectrometer (TOMS) experiment. *In* "Nimbus 7 User's Guide." Goddard Space Flight Cent., Greenbelt, Maryland.

Hovis, W. (1978). The coastal zone color scanner (CZCS) experiment. *In* "Nimbus 7 User's Guide," pp. 19–31. Goddard Space Flight Cent., Greenbelt, Maryland.

McPeters, R. D., Heath, D. F., and Bhartin, P. K. (1984). Average ozone profiles for the Nimbus 7 SBUV instrument. *JGR, J. Geophys. Res.* **89**(D4), 5199–5214.

Morse, P. G. (1980). Progress report on the ATMOS sensor: design description and development status. *AIAA Pap.* **80-1914,** 13–24.

Nagatani, R. M., and Miller, A. J. (1984). Stratospheric ozone changes during the first year of SBUV observations. *JGR, J. Geophys. Res.* **89**(D4), 5191–5198.

Roche, A. E., James J. B., Kumer, T. C., Nast, T. C., and Sears, R. D. (1979). A cryogenic etalon spectrometer for measurement of stratospheric and mesospheric minor species and temperatures from the Space Shuttle. *IEEE Space Instrum. Atmos. Observ. Conf.* pp. 158–168.

Roche, A. E., Forney, P. B., Kumer, J. B., Naes, L. G., and Nast, T. C. (1982). Performance analysis for the cryogenic etalon spectrometer on the upper atmosphere research satellite. *Proc. Soc. Photo-Opt. Instrum. Eng.* **364,** 46–58.

Smith, W. L., Howell, H. B., and Woolf, H. M. (1979). The use of interferometric radiance measurements for sounding the atmosphere. *J. Atmos. Sci.* **36,** 566–575.

Smith, W. L., Revercomb, H. E., Howell, H. R., and Woolf, H. M. (1983). HIS—a satellite instrument to observe temperature and moisture profiles with high vertical resolution. *Conf. Atmos. Radiat. 5th, Am. Meteorol. Soc.* pp. 1–9.

Thuillier, J. F., Brun, J. F., Monge, J. L., Sirou, F., Semery, A., Duboin, M. L., Porteneuve, J., and Chandra, S. (1981). Instrument requirements document for the temperature and wind measurement in the mesosphere and the thermosphere. *In* "Upper Atmosphere Research Satellites." Goddard Space Flight Cent., Greenbelt, Maryland.

Twomey, S. (1961). On the deduction of the vertical distribution of ozone by ultraviolet spectra measurements from a satellite. *J. Geophys. Res.* **66,** 2153–2162.

Vanasse, G. A., ed. (1977). "Spectrometric Techniques," Vol. I. Academic Press, New York.

Viollier, M., and Sturm, B. (1984). CZCS data analysis in turbid coastal water. *JGR, J. Geophys. Res.* **89**(D4), 4977–4987.

Wark, D. Q. (1960). Doppler widths of the atomic oxygen lines in the airglow. *Astrophys. J.* **131,** 491–501.

Wark, D. Q., and Hilleary, D. T. (1969). Atmospheric temperature: successful test of remote probing. *Science (Washington, D.C.)* **165,** 1256–1258.

Wark, D. Q., Hillary, D., Lienesch, J., and Clark, P. (1969). The satellite infrared spectrometer (SIRS) experiment. *In* "Nimbus 3 User's Guide," pp. 147–179. Goddard Space Flight Cent., Greenbelt, Maryland.

Wellman, J. B., Goetz, A. F. H., Herring, M., and Vane, G. (1983). An imaging spectrometer experiment for the Shuttle. *IEEE Int. Geosci. Remote Sens. Symp.* **83CH1837-4,** 6.1–6.7.

Chapter 6 | Passive Space Microwave Radiometer Systems

6.1 Introduction

Passive microwave space sensors measure microwave radiation emitted from the earth's surface and atmosphere. The signal output from the sensor can be used to generate the surface temperature, vertical temperature profile, and atmospheric species concentrations. Passive space remote sensing in the microwave spectral region has applications that range from earth environmental studies to oceanography. Each region of the electromagnetic spectrum has unique properties that can be exploited for space remote sensing, and in the microwave region these properties include the ability to penetrate clouds and to study absorption by O_2, H_2O, and other atmospheric constituents.

The main advantage of a space microwave radiometer over a space infrared radiometer is the much reduced effect of clouds at microwave frequencies. The disadvantages are the microwave antenna size and weight, the large field of view (FOV), and the smaller amount of emitted radiation. For these reasons microwave radiometers will never replace infrared radiometers, and the best system may be one that combines an infrared sensor with a microwave sensor.

The microwave electromagnetic spectrum that will be used in space remote sensing applications is shown in Table 6.1 as a function of wavelength and frequency. The nomenclature, symbolism, and units used to express quantities in microwave radiometry that are different from those used in optical radiometry are listed in Table 6.2.

According to Planck's radiation equation, a blackbody radiates uniformly in all directions with a spectral brightness B_f given by

$$B_f = \frac{2hf^3}{c^2} \frac{1}{\exp(hf/kT) - 1} \tag{6.1}$$

where B_f is the blackbody spectral brightness in watts per square centimeter steradian hertz, h Planck's constant, f the frequency in hertz, k Boltzmann's constant, T the absolute temperature in degrees kelvin, and c the velocity of light.

From the above equation, the spectral brightness is a function of f and T. Figure 1.2 shows spectral brightness as a function of temperature and frequency for blackbody radiation. From the figure it is clear that the earth emits microwave radiation as well as infrared radiation, and the microwave radiation from the earth and its atmosphere can be used as the source of radiation for passive space remote sensing applications.

The radiative transfer equation in the microwave region for atmospheric sounding and surface imaging can be written in terms of the brightness, which corresponds to the radiance in the infrared region. The upwelling microwave brightness can be expressed as

$$B_{\mathrm{m}} = \varepsilon B(f, T_{\mathrm{s}})\tau(f, P_{\mathrm{s}}) + (1 - \varepsilon)\tau(f, P_{\mathrm{s}}) \int_1^{\tau(f, P_{\mathrm{s}})} B(f, P)\, d\tau(f, P) - \int_1^{\tau(f, P_{\mathrm{s}})} B(f, P)\, d\tau(f, P) \tag{6.2}$$

Table 6.1

Microwave Spectra Definitions

Region	Wavelength	Frequency (GHz)
Microwave	1 m–1 cm	0.3–30
Millimeter	1 cm–0.1 cm	30–300
Submillimeter	0.1 cm–0.01 cm	300–3000

Table 6.2

Fundamental Radiometric Parameters

Optical terminology	Microwave terminology	Symbol	Unit
Radiant flux	Power	P	W
Irradiance	Power density	S	W/cm^2
Radiance	Brightness	B	W/cm^2 sr
Spectral radiance (Sr)	Spectral brightness (Sb)	B	W/cm^2 sr Hz (Sb) W/cm^2 sr μm (Sr)

where B_m is the microwave radiation brightness, $B(f, T, P)$ the Planck function, f the microwave frequency, τ the transmittance, T the temperature, P the atmospheric pressure, ε the emissivity of the earth's surface, and P_s the surface pressure.

The first term on the right-hand side of this equation denotes the surface microwave emission. The second term represents the microwave emission from the entire atmosphere to the earth's surface, which is reflected back to the atmosphere at the same frequency. The third term denotes the upwelling microwave emission by the entire atmosphere.

6.2 Microwave Sensor Spectral Band Selection

Microwave space remote sensing applications can be divided into thermal imaging and atmospheric sounding. Thermal imaging sensors operate primarily at window frequencies, where atmospheric species absorption is low and the temperature or thermal images of the surface can be measured. Atmospheric sounders provide information about vertical profiles of temperature and species concentration by sensing the emitted radiation at different heights in the atmosphere.

6.2.1 *Microwave Thermal Imaging Spectral Band Selection*

Microwave spectral band selection can be divided into three main categories: land, ocean, and ice. Atmospheric radiation at these microwave frequencies is affected to some extent by water vapor and rainfall. Hence most remote sensing passive radiometer systems include spectral frequency bands that are sensitive to water vapor to correct for their effects on the measurement of surface temperature or soil moisture. In addition, the spectral bands are selected so that the systems can operate in regions of the spectrum free of interference from ground radars and other communication systems. Table 6.3 shows the spectral band selection for microwave imaging.

6.2.2 *Atmospheric Sounding Spectral Band Selection*

The Nimbus E (Nimbus 5) microwave spectrometer (NEMS) was the first space sensor for atmospheric sounding in the microwave region. The mixing ratio of atmospheric oxygen is quite uniform; thus measurements of atmospheric emission at frequencies around the oxygen band centered near 60 GHz can be used to determine the vertical temperature at altitude

Table 6.3

Microwave Imaging Spectral Band Selection

Frequency (GHz)	Application
1.4	Surface soil moisture
4.3	Sea surface temperature
5.2	Sea surface temperature
6.6	Sea surface temperature
10.4	Sea surface temperature
10.7	Surface wind field
13.9	Surface roughness
19.3	Surface features
21.0	Atmospheric water content
22.2	Atmospheric water content
31.4	Surface imaging
35.0	Surface imaging
37.0	Surface ice, oil
90.0	Surface imaging
135.0	Surface imaging
180.0	Atmospheric water content

levels defined by temperature weighting functions. By choosing different frequencies whose weighting functions peak at different altitudes, retrieval techniques can be used to determine vertical temperature profiles by inversion of the microwave remote sensing data. Table 6.4 shows the spectral channel selection for temperature sounding.

Millimeter-wave space sensors can be used to measure thermal limb atmospheric emission in certain spectral regions to obtain upper-atmosphere observations for vertical species and temperature profile determination. These limb remote sensing techniques can provide substantially improved sensitivities and vertical resolution. Table 6.5 shows the spectral band selection for limb sounding.

6.3 Microwave and Millimeter-Wave Sensor Antenna Systems

Passive microwave and millimeter-wave sensor systems consist of two types of radiometers: wideband radiometers and narrowband multispectral radiometers. The wideband radiometers are used to measure temperature, ocean wind speed, ice pack, rain rate, water content, and temperature profile. The narrowband radiometers are used to measure molecular

Table 6.4

Microwave Nadir Sounding Spectral Band Selection

Frequency (GHz)	Application
50.3	Temperature sounding
52.8	
53.2	
53.8	
54.3	
54.9	
57.9	
58.4	
58.8	
59.4	

line profiles, from which atmospheric species concentration, temperature, pressure, and wind can be determined. Microwave space sensors consist of a collecting antenna, a heterodyne receiver, and associated rf and digital electronics for data analysis. The heterodyne receiver is composed of a local oscillator, a mixer, preamplifiers, and related electronic systems. For spectroradiometer applications a spectral line receiver must also be used to provide the power spectrum of the mixer intermediate frequency (IF) output.

The antenna diameter of the micowave sensor can be determined by the spatial resolution, wavelength of the spectral band, and space sensor

Table 6.5

Limb Sounding Spectral Band Selection

Frequency (GHz)	Applications
63.56	Pressure
118.75	O_2 vertical profiles
119.00	Temperature and wind
183.31	H_2O vertical profiles
204.35	ClO and O_3 vertical profiles
204.57	H_2O_2 vertical profiles
206.13	O_3 vertical profiles
230.53	CO vertical profiles

height, as shown in the following equation:

$$D = \text{altitude} \times \frac{1.2}{\text{spatial resolution}} \tag{6.3}$$

The sensor antenna can be of either the parabolic or the Cassegrain type. The antenna reflector can be constructed by using a rib-supported honeycomb shell made of an aluminum honeycomb core and graphite/epoxy face skins. The thermal profile of the antenna reflector is controlled passively by using a thermal coating on the front face and insulation on the back

The parabolic antenna system has better beam efficiency because of the clear aperture, spillover that mostly observes the cold space background, which introduces small measurement errors, and a single reflector, which reduces the cost. The Cassegrain antenna system has a secondary reflector, which reduces the beam efficiency, forward spillover that observes the earth's radiation, which introduces significant measurement errors, and a two-reflector system, which increases the cost.

With regard to space remote sensing, it appears certain that antennas of the order of 100 m in diameter operating in the 10–20-cm microwave band will produce ground resolution of 1 km from low earth orbit in the 1990s and beyond. A new concept consists of using a linear radiometer array perpendicular to the orbital path of the spacecraft. The advantage of this pushbroom concept is that no mechanical scanning of the antenna is required. The time delay and integration (TDI) technique can also be used for better signal collection. Figure 6.1 shows antenna size versus spatial resolution.

6.4 Microwave Radiometer System

The most useful passive space microwave radiometer system is the Dicke superheterodyne type. A Dicke switch changes periodically at a high rate between the incoming signal and a known reference source. The local oscillator (LO) frequency is at the center of the desired radiometer bandwidth and is mixed with the input signal to produce an IF signal, which is then amplified and detected by a square-law detector and a synchronous detector. The signal then passes through the integrator, multiplexer, and analog-to-digital (A/D) converter, and is transmitted to the ground station.

In addition to the Dicke superheterodyne system, there is another type of passive microwave radiometer system: the total power superheterodyne type. The Dicke type is basically a total power type with two special

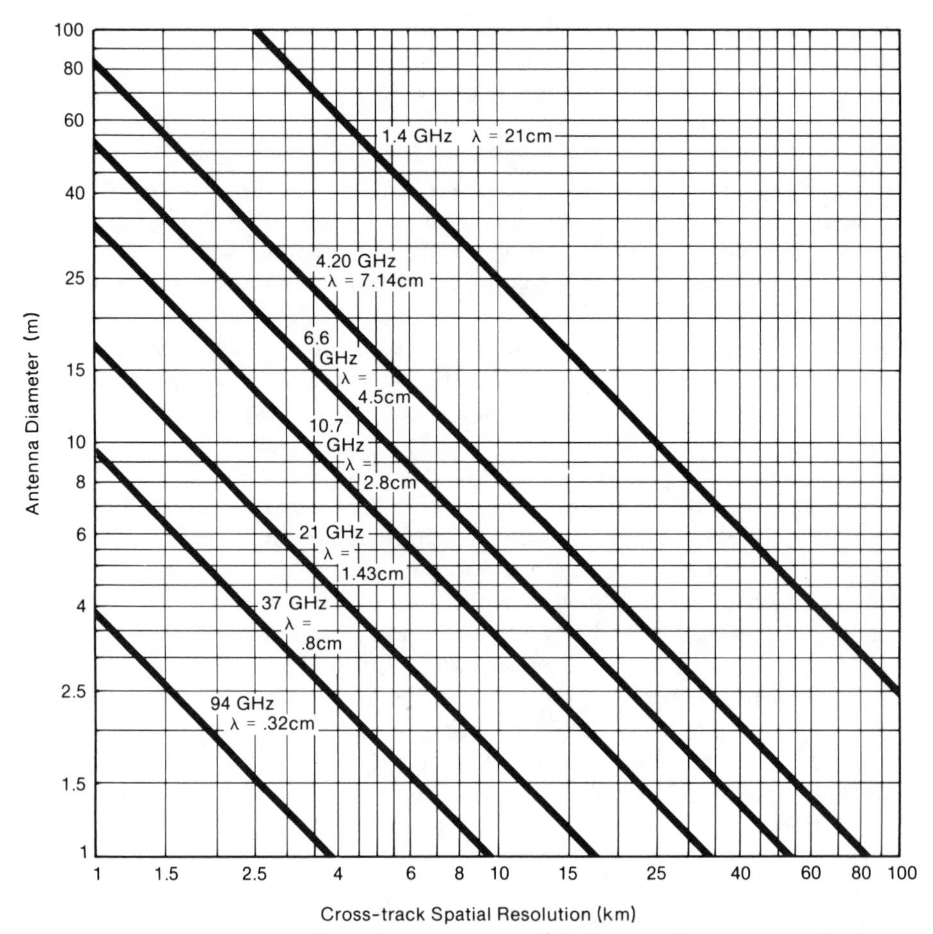

Fig. 6.1. Antenna size versus resolution for 600 km altitude sensor with 45° nadir scanning angle.

additional subsystems: the Dicke switch and the synchronous detector. The radiometric resolution of the Dicke radiometer is superior to that of the total power radiometer by a factor of 2 to 50 in the range of antenna temperatures from 0 to 300 K.

The microwave radiometric sensitivity or resolution is regarded as the minimum detectable change in the radiometric antenna temperature T of the observed earth-emitted microwave radiation. The radiometric sensitivity of a simple total power radiometer is given by the following equation:

$$\Delta T = (T_{\mathrm{A}} + T_{\mathrm{RC}})[1/Bt + (\Delta G/G)^2]^{1/2} \tag{6.4}$$

where ΔT is the radiometer sensitivity, T_A the antenna noise temperature, T_{RC} the receiver input noise temperature, B the receiver bandwidth, t the integration time, ΔG the power gain variation, and G the average system power gain.

The radiometric sensitivity of a Dicke superheterodyne radiometer is given by

$$\Delta T = \{[2(T_A + T_{RC})^2 + 2(T_{RC} + T_{RF})^2]/Bt + (\Delta G/G)^2(T_A - T_{RF})^2\}^{1/2} \tag{6.5}$$

where T_{RF} is the reference source noise temperature. If $T_A = T_{RF}$, the radiometer is in a balanced condition. In this case the radiometric sensitivity is given by

$$\Delta T = 2(T_A + T_{RC})/(Bt)^{1/2} \tag{6.6}$$

In space remote sensing of the earth the radiometer will see an antenna temperature of approximately 300 K. For a noise receiver at 300 K with 1-sec integration time and a bandwidth of 100 MHz, ΔT is equal to 0.12 K. If 1 K radiometer sensitivity is required, then 0.12-sec integration time can be used for the radiometer system.

A system block diagram of a Dicke-switched superhetrodyne radiometer is shown in Fig. 6.2.

6.5 Microwave Antenna Mass Determination

Space antenna technology is very important for microwave radiometer systems. The spatial resolution is determined by the antenna dimensions and increases with antenna size. The antenna reflectors can be of fixed dimensions or deployable to larger size after being launched into space. Thermal and mechanical variations may require that antenna structural deflection be sensed and controlled actively if passive methods are not adequate.

One of the greatest concerns in space antenna design is the size and mass of the antenna structure. Various approaches for deployable antennas are being investigated, including flex-rib or ridged-rib/mesh, expandable truss/mesh, and multisegmented surfaces. Fixed-size antenna technology is directed toward the development of techniques for producing precise surfaces that can be used in a space environment.

Precision microwave graphite/epoxy reflectors have been developed for space applications in different sizes, with diameters in the range of a few meters and surface tolerances varying from 0.0001 to 0.010 cm rms. Diameters of fixed antennas are limited by the Space Transportation Sys-

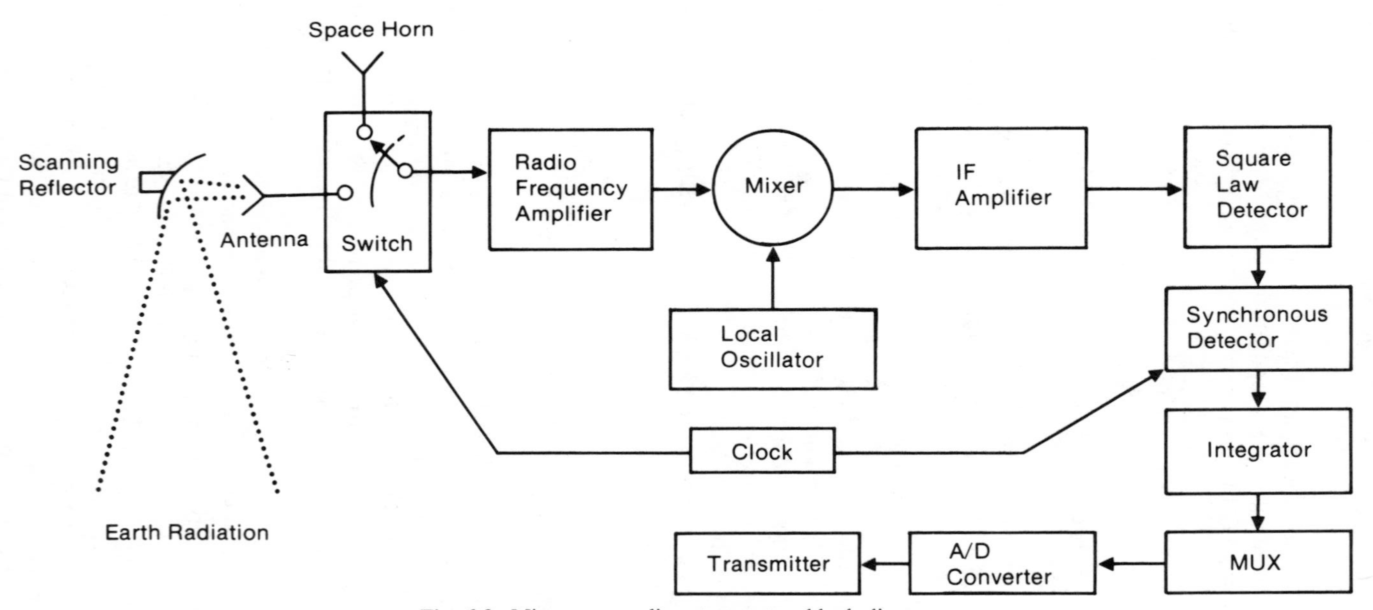

Fig. 6.2. Microwave radiometer system block diagram.

Table 6.6

Weight of Single Parabolic Reflector versus Phased-Array Antenna

Antenna diameter (m)	Antenna weight (kg)	
	Parabolic reflector	Phased array
10	80–100	40–70
20	300–400	150–250
40	900–1200	400–550
60	1400–1700	700–900
100	3000–4000	1200–1600
120	5000–7000	2000–3000

tem (STS) payload bay envelope, which is 4.3 m in diameter. Deployable antennas are expected to reach 100 m by the year 2000.

Table 6.6 shows the microwave antenna weight as a function of diameter for a fixed parabolic reflector system and a phased-array system. The parabolic system, which needs more structure to achieve the required surface curvature and tolerances, will be a heavier space reflector system.

6.6 Space Microwave Sensors and Applications

6.6.1 *Microwave Imagers and Applications*

6.6.1.1 *The Electrically Scanning Microwave Radiometer (ESMR)*

In December 1972, the ESMR was launched aboard Nimbus 5 for surface mapping in the 19.35-GHz channel from space. The ESMR consisted of four major subsystems: a phased-array microwave antenna consisting of 103 waveguide elements, each with associated electrical phase shifters; a beam-steering computer, which determined the coil current for each of the phase shifters for each beam position; a microwave receiver with a center frequency of 19.35 GHz; and timing control and power circuits. The aperture size of the phased-array antenna was 83.3 × 85.5 cm, with polarization linear and parallel to the spacecraft velocity vector. The antenna beam was scanned cross-track in 78 steps, ±50° from nadir, within 4 sec. Scanning was achieved by means of the phase shifters in series with each waveguide element and controlled by timing commands

Table 6.7

ESMR Sensor Parameters

Frequency, GHz	19.35
Wavelength, cm	1.55
rf bandwidth, MHz	250
Integration time, sec	0.047
Absolute accuracy, K	2.0
Dynamic range, K	50–330
IF frequency, MHz	5–125
Antenna beamwidth, deg	1.4 × 1.4, nadir 2.2 × 1.4, end of scan
Antenna beam efficiency, %	90–92.7

via the beam-steering network. The antenna beamwidth was 1.4 × 1.4° near nadir and 2.0 × 1.4° at the 50° extremes. From an orbit of 1100-km altitude the footprint resolution was 25 × 25 km in the nadir direction and degraded to 160 × 45 km at the end of the scan. The swath width generated by the cross-track scan was 1570 km for global coverage within 12 hr.

The ESMR has found important applications in research on polar ice caps, detecting precipitation, and determining soil moisture conditions. The microwave radiometer was of the Dicke-switched superheterodyne type with two calibration reference sources. One calibration reference source was provided by a matched waveguide termination at ambient temperature. The other was provided by a sky horn, which viewed the 3 K space background temperature. By means of these two calibration points and the known losses in the waveguide and switch system, the microwave sensor system could be calibrated to 2 K absolute accuracy. Table 6.7 lists the ESMR sensor parameters.

6.6.1.2 *Nimbus 6 Electrically Scanning Microwave Radiometer (N6ESMR)*

The N6ESMR sensor was an upgraded version of the Nimbus 5 ESMR microwave radiometer. It collected thermal microwave radiation from the earth's atmosphere and surface system in a spectral channel centered at 37 GHz (0.81 cm). This frequency change from 19.35 GHz (1.55 cm) to 37 GHz had the effect of tripling the sensor's sensitivity to water droplets in measurements for distinguishing areas of light rain from areas of high water vapor, and of doubling the contrast between first-year ice and multiyear ice. In general, the sensor data will be used to map the liquid-water content of the clouds, the distribution and variation of sea ice cover and snow cover on the ice, and the condition of land surfaces.

The sensor measured both vertically and horizontally polarized earth radiation at the 37-GHz frequency, as opposed to horizontal only for the Nimbus 5 ESMR. The antenna beam scanned ahead of the spacecraft along a conical surface at a constant angle of 45° with respect to the antenna axis. The beam scanned in azimuth ±35° about the forward direction in 71 steps. The resolution field of view (RFOV) on the earth's surface was nearly 20 × 43 km. The scan pattern resulted in a swath width of 1300 km. Atmospheric radiation of both polarizations was received and separated in the phased-array antenna before being fed into the radiometer. The dual-polarization capability of the N6ESMR was used to study its effectiveness for improved measurements of wind speed over the oceans and rainfall rate over the land and the oceans. The radiometer section of the instrument was similar in operation to that of the Nimbus 5 ESMR. Figure 6.3 shows Nimbus 6 and its ESMR.

6.6.1.3 *The Scanning Multichannel Microwave Radiometer* (*SMMR*)

The scanning multichannel microwave radiometer is a five-channel imaging microwave radiometer system. It was launched on Nimbus 7 and the Seasat satellites. The sensor measures dual-polarized radiation from the earth at frequencies of 6.6, 10.7, 18, 21, and 37 GHz. The 6.6-GHz channel was chosen for sea surface temperature measurements, the 10.69-GHz channel for wind-field observation, and the 18-, 21-, and 37-GHz channels to provide additional data on atmospheric water vapor, cloud liquid water, and rainfall. Polarization is used in all channels to provide additional discrimination for better data analysis.

The SMMR consists of six Dicke superheterodyne radiometers, fed by a single antenna and a calibration subsystem. Two radiometers simultaneously measure the polarization components of the received earth radiation. The other four radiometers alternate polarization components during successive scans. The SMMR has a scanning antenna system. The antenna is an offset parabolic reflector with a 79-cm aperture. The antenna reflector is mechanically scanned about a vertical axis over a ±25° azimuth angle range. The antenna FOV is offset 42° from nadir and thus sweeps out the surface of a cone and provides a constant viewing angle at the earth's surface. A data programmer unit in the electronic system provides the timing, multiplexing, and synchronization signals, contains A/D converters, multiplexers, and shift registers, and provides formatting and buffering functions between the sensor and the spacecraft digital system. Table 6.8 lists the SMMR sensor parameters, and Fig. 6.4 shows the SMMR sensor configuration.

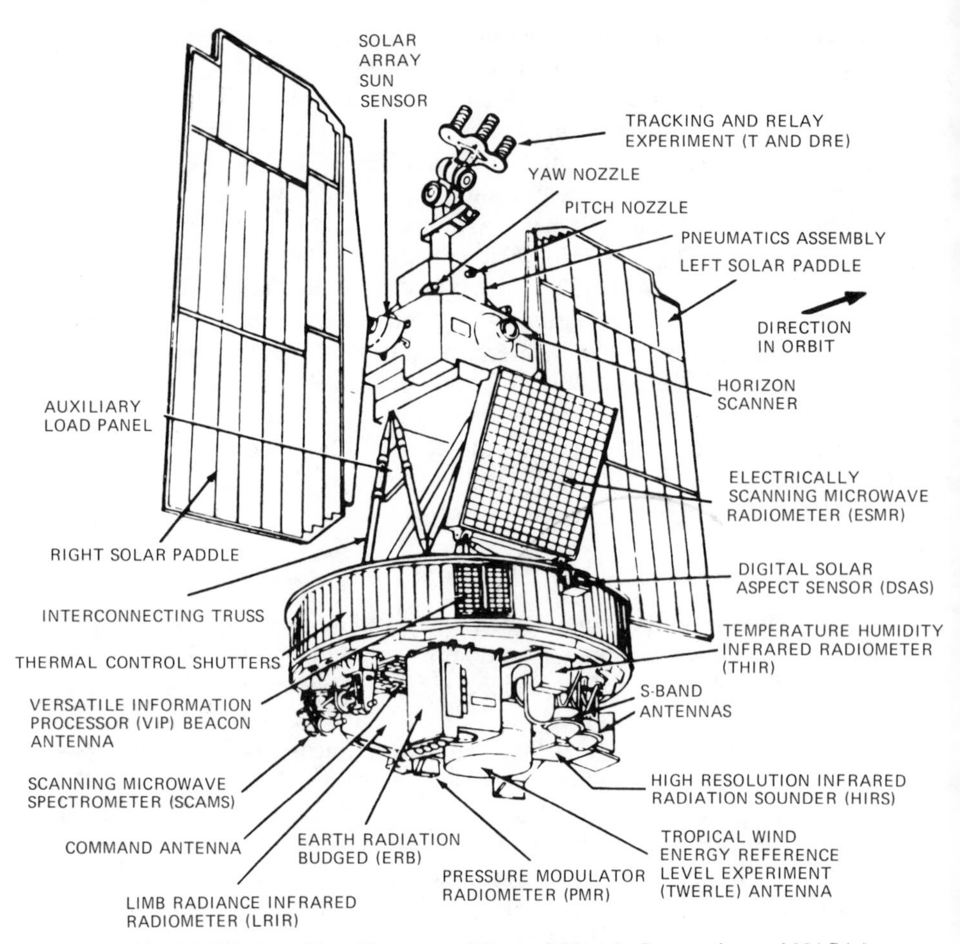

Fig. 6.3. Nimbus 6 satellite. (Courtesy of General Electric Corporation and NASA.)

6.6.2 *Microwave Sounders and Applications*

6.6.2.1 *The Nimbus E Microwave Spectrometer (NEMS)*

The NEMS (N5MS) was the first passive microwave temperature sounder for space remote sensing applications. The sensor consisted of three types of space instruments used for measuring vertical atmospheric temperature profiles, surface temperature, and water vapor and liquid water over the oceans. The oxygen band radiometer consisted of channels at 53.65, 54.9, and 58.8 GHz for temperature sounding applications. The

Table 6.8

Sensor Parameters

Parameters	SMMR	NEMS	SCAMS
Frequency	6.6	22.235	22.235
	10.69	31.4	31.65
	18.00	53.65	52.85
	21.00	54.9	53.85
	37.00	58.8	55.45
rf bandwidth, MHz	250	250	220
Integration time, sec	0.12	2.0	0.95
Sensitivity, K	0.9–1.5	0.23–0.29	0.2–0.6
Dynamic range, K	10–330	100–325	0–350
Absolute accuracy, K	2.0	2.0	2.0
IF frequency range, MHz	10–110	10–100	10–100
Antenna beamwidth, deg	4.2	10	7.5
Antenna beam efficiency, %	87.0	93–95	96.9–97.4

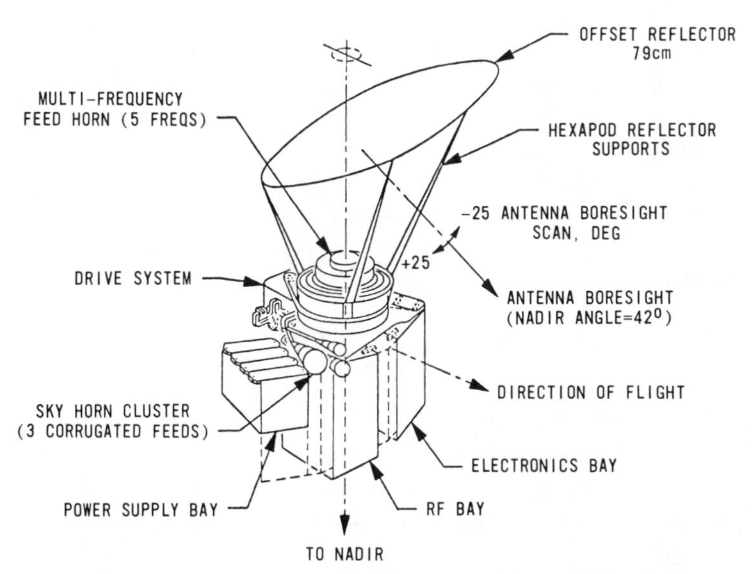

Fig. 6.4. SMMR sensor configuration. (Courtesy of Jet Propulsion Laboratory and NASA.)

water vapor radiometer consisted of a 22.24-GHz channel for water vapor measurements. The atmospheric window radiometer consisted of a 31.40-GHz channel for surface temperature measurements.

Each of the five microwave radiometer bands was of the Dicke-switched superheterodyne type, with a two-point calibration system. The input and calibration switches allowed a selection of inputs to the instrument from either the antenna, the space low-temperature calibration, or the instrument ambient temperature. The receiver portion of the channel was a conventional superheterodyne system isolated from the Dicke switch. The signal was routed to a low-noise balanced mixer with a solid-state local oscillator. An IF amplifier, detector, video amplifier, and dc amplifier completed the radiometer system. The dc output was connected to the data processing unit, where it was processed prior to entering the data stream as prime data. The signal also passed through a buffer circuit, and the output was used as an analog signal for quick-look and prime data backup capability. The NEMS sensor parameters are listed in Table 6.8.

6.6.2.2 *The Scanning Microwave Spectrometer (SCAMS)*

The SCAMS on Nimbus 6 was an upgraded sensor that had the same objective as the NEMS sensor on Nimbus 5. The NEMS measured atmospheric parameters in the nadir direction. The SCAMS operated by scanning to either side of the subsatellite track to produce nearly full earth coverage every 12 hr. Sensor scanning was achieved by stepping a radiation reflector in front of each antenna. Each reflector scanned the earth radiation and a completed 360° scan took 16 sec. There were 13 radiation data samplings within ±43.2° from nadir, each separated by a 7.2° scan step. The antenna beamwidth was 7.5°, which gave a ground resolution of 145 km at nadir and 220 × 360 km at the edge of the scan. The complete scan covered a swath width of 2400 km.

Weighting functions for the three temperature sounding channels are shown in Fig. 6.5. Table 6.8 lists the sensor parameters.

6.6.2.3 *Microwave Sounding Unit (MSU)*

The microwave sounding unit is one of the payloads of the TIROS-N/NOAA operational satellite series. The MSU is an adaptation of the SCAMS sensor flown on Nimbus 6. It is a four-channel Dicke radiometer that makes passive microwave radiation measurements in four regions of the 60-GHz or 5.5-mm oxygen absorption band for vertical atmospheric temperature sounding.

The MSU has two scanning reflector antenna systems. The antennas scan ±47.4° on either side of nadir in 11 steps. The antenna beamwidth is

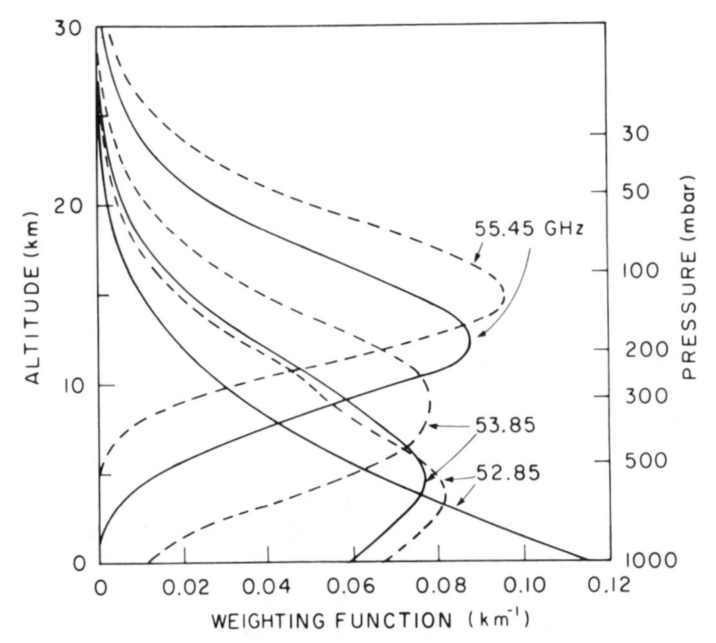

Fig. 6.5. SCAMS weighting functions: dashed line: maximum scan angle; solid line: nadir. (Courtesy of D. H. Staelin and NASA.)

7.5°, resulting in a sampling resolution at the subsatellite point of 109 km. Atmospheric microwave radiation received by each antenna is separated into vertical and horizontal polarization components by an orthomode transducer. Each of the four radiation signals is fed to one of the radiometer channels. The incoming radiation is modulated at 1 kHz by a Dicke switch so that a constant comparison is made between the ambient temperature source and the incoming radiation.

Table 6.9 lists the MSU sensor system parameters.

6.6.2.4 *Microwave Limb Sounder* (*MLS*)

The microwave limb sounder is one of the payloads of the upper atmosphere research satellite (UARS). The MLS sensor objective is to improve our understanding of the earth's upper atmosphere by measuring vertical atmospheric species concentrations and temperature profiles, atmospheric wind, and magnetic field.

The MLS consists of two major subsystems: a sensor whose FOV is scanned through the atmospheric limb to measure vertical profiles of atmospheric parameters, and an electronics system. The major elements

Table 6.9

MSU Sensor Parameters[a]

Parameters	Channel 1	Channel 2	Channel 3	Channel 4
Frequency, GHz	50.3	53.74	54.96	57.05
rf bandwidth, MHz	220	220	220	220
Sensitivity, K	0.3	0.3	0.3	0.3
Dynamic range, K	0–350	0–350	0–350	0–350
Antenna beam efficiency, %	90	90	90	90

[a] Courtesy of NOAA.

of the sensor are the antenna system, which defines the FOV, calibration targets and switching mirrors, a multiplexer that separates the signal from the switching mirror into five spectral bands centered at 63, 119, 183, 205, and 233 GHz. The electronics system includes a filter bank and an autocorrelator spectrometer, which perform spectral analysis of the signals from the radiometers, a data control system, and a power supply.

Ozone vertical profiles are measured by the 206.13- and 235.70-GHz spectral lines. Both ClO and H_2O_2 are simultaneously measured by the 204.352- and 204.546-GHz emission lines. The CO vertical profile is measured at 230.538 GHz. The H_2O profile is measured by remote limb sensing at 183.31 GHz. The O_2 vertical profile is measured by the 119-GHz line, and the vertical temperature profile is also measured by this line. Winds can be measured by the 119-GHz line due to magnetic dipole

Table 6.10

MLS Measurement Objective[a]

Parameters	Altitude range (km)	Vertical resolution (km)
O_3	15–90	3
ClO	25–40	3
H_2O_2	30–40	3
CO	20–100	3
H_2O	25–90	4
O_2	80–120	6
Temperature	25–100	6
Magnetic field	80–100	6

[a] Courtesy of NASA.

transitions and the splitting of the line into its Zeeman components by the earth's magnetic field. Table 6.10 shows the MLS measurement objectives.

General References and Bibliography

Allan, T. D., ed. (1983). "Satellite Microwave Remote Sensing." New York Univ. Press, New York.

Bernstein, R. L. (1982). Sea surface temperature mapping with the Seasat microwave radiometer. *JGR, J. Geophys. Res.* **87,** 7865–7872.

Chang, H. D., Hwang, P. H., Wilheit, T. T., Chang, W. J., Staelin, D. H., and Rosenkranz, P. W. (1984). Monthly distribution of precipitable water from the NIMBUS SMMR data. *JGR, J. Geophys. Res.* **89,** 5328–5334.

Dicke, R. H., Beringer, R., Kyhl, R. L, and Vans, A. B. (1946). Atmospheric absorption measurements with a microwave radiometer. *Phys. Rev.* **70,** 340–348.

Gloersen, P., and Barath, F. T. (1977). A scanning multichannel microwave radiometer for Nimbus-G and Seasat-A. *IEEE J. Oceanic Eng.* **OE-2,** 172–200.

Gloersen, P., and Hardis, L. (1978). The scanning multichannel microwave radiometer (SMMR) experiment. *In* "The Nimbus 7 User's Guide," pp. 213–245. Goddard Space Flight Cent., Greenbelt, Maryland.

Gloersen, P., Wilheit, T. T., Chang, T. C., and Nordberg, W. (1974). Microwave maps of the polar ice of the earth. *Bull. Am. Meteorol. Soc.* **55,** 1442–1448.

Grody, N. C. (1976). Remote sensing of atmospheris water content from satellites using microwave radiometry. *IEEE Trans. Antennas Propag.* **AP-24,** 155–162.

Hollinger, J. P., and Lo, R. C. (1984). Low frequency microwave radiometer. *Proc. Soc. Photo-Opt. Instrum. Eng.* **481,** 199–207.

Njoku, E. G. (1982). Passive microwave remote sensing of the earth from space—A review. *Proc. IEEE* **70,** 728–750.

Njoku, E. G., Stacey, J. M., and Barath, F. T. (1980). The Seasat scanning multichannel microwave radiometer (SMMR): instrument description and performance. *IEEE J. Oceanic Eng.* **OE-5,** 100–115.

Staelin, D. H. (1969). Passive remote sensing at microwave wavelengths. *Proc. IEEE* **57,** 427–439.

Staelin, D. H., Barath, F. T., Blinn, J. C., and Johnson, E. J. (1972). The Nimbus 5 microwave spectrometer (NEMS) experiment. *In* "The Nimbus 5 User's Guide," pp. 141–157. Goddard Space Flight Cent., Greenbelt, Maryland.

Staelin, D. H., Barrett, A. H., Rosenkranz, P. W., Barath, F. T., Johnson, E. J., Waters, J. W., Wouters, A., and Lenoir, W. B. (1975). The scanning microwave spectrometer (SCAMS) experiment. *In* "The Nimbus 6 User's Guide," pp. 59–86. Goddard Space Flight Cent., Greenbelt, Maryland.

Tomiyasu, K. (1974). Remote sensing of the earth by microwave. *Proc. IEEE* **62,** 86–92.

Ulaby, F. T., Moore, R. K., and Fung, A. K. (1981). "Microwave Remote Sensing: Active and Passive." Addison-Wesley, Reading, Massachusetts.

Water, J. (1981). Instrument requirements document for the microwave limb sounder. *In* "Upper Atmosphere Research Satellites." Goddard Space Flight Cent., Greenbelt, Maryland.

Water, J., Kunzi, K. F., Pettyjohn, R. L., Poon, R., and Staelin, D. H. (1975). Remote sensing of atmospheric temperature profiles with the Nimbus-5 microwave spectrometer. *J. Atmos. Sci.* **32,** 1953–1969.

Wilheit, T. (1972). The electrically scanning microwave radiometer (ESMR) experiment. *In* "The Nimbus 5 User's Guide," pp. 59–105. Goddard Space Flight Cent., Greenbelt, Maryland.

Wilheit, T. (1975). The electrically scanning microwave radiometer (ESMR) experiment. *In* "The Nimbus 6 User's Guide," pp. 87–108. Goddard Space Flight Cent., Greenbelt, Maryland.

Chapter 7 Active Space Lidar Systems

7.1 Introduction

Passive space sensors have been developed for space remote sensing for many years. Active laser systems on an orbiting satellite will provide new capabilities for high-specificity and high-spatial-resolution measurements in day and night operation. A lidar (light detection and ranging) sensor consists of lasers, an optical system, detectors, electronics, and a data recording system. The active system is really a combination of a special light source system and a passive sensor system. The space laser is usually operated in a pulsed mode. The laser beam interacts with the target atmosphere and the earth's surface and the backscattered laser radiation is collected by the optical system and measured by the detector and electronics systems.

The Space Shuttle or Space Station would provide the ideal platform for lidar experiment missions designed to contribute to an understanding of the processes governing the earth's atmosphere. Space lidar systems can make an important contribution in several major scientific applications, including remote sensing of atmospheric species, high-resolution radiative transfer modeling, cloud physics measurements, high-resolution vertical temperature profiles, and surface albedo determinations.

Space lidar sensors can be divided into two broad classes: direct and heterodyne sensor systems. In direct sensor systems, signal current is generated in direct proportion to received laser power; that is, signal response is processed by square-law-type detectors such as photomultiplier tube (PMT), photodiode, or photovoltaic (PV) and photoconductive

(PC) detectors. Frequency and phase information is lost in the detection process. In heterodyne systems, the mixer uses a local oscillator to generate a signal at the difference IF between the local oscillator and the received signal. Heterodyne lidar systems have superior sensitivity and very high spectral resolution, but are complicated by the use of a secondary laser source and alignment problems. The direct lidar sensor is a relatively simple, reliable, and low-cost system. The sensitivity of the lidar heterodyne system is three orders better than that of the direct system. Figure 7.1 shows the configuration of lidar direct and heterodyne systems.

7.2 Space Lidar Spectral Line Selections

7.2.1 *Carbon Dioxide Laser Spectral Line Selection*

Carbon dioxide laser technology has been extensively developed for remote sensing applications. The CO_2 laser can operate on dozens of lines in the 9–12-μm range at a low pressure of 1 atm. Overall power conversion efficiencies are typically 10% for CO_2 lasers. There should be no major problem in applying either a pulsed or a cw CO_2 laser in space. Table 7.1 list the laser line selection for space remote sensing applications.

7.2.2 *Nd : YAG Laser Spectral Line Selection*

The Nd : YAG laser has been extensively developed for remote sensing applications. It can operate with about 1% overall efficiency at 1.06 μm and about half that efficiency in the double-frequency mode. It can also be frequency-tripled and quadrupled to produce useful outputs in the ultraviolet, and it can be used to pump a dye laser in the visible and UV spectral regions. There should be no difficulty using this laser in space applications. Table 7.2 lists Nd : YAG laser spectral lines for various applications.

7.2.3 *Alexandrite Tunable Solid-State Laser Line Selection*

New solid-state laser materials such as chromium-doped alexandrite make very promising tunable lasers for remote sensing applications. The spectral output of the alexandrite laser is in the 700–800-nm region and

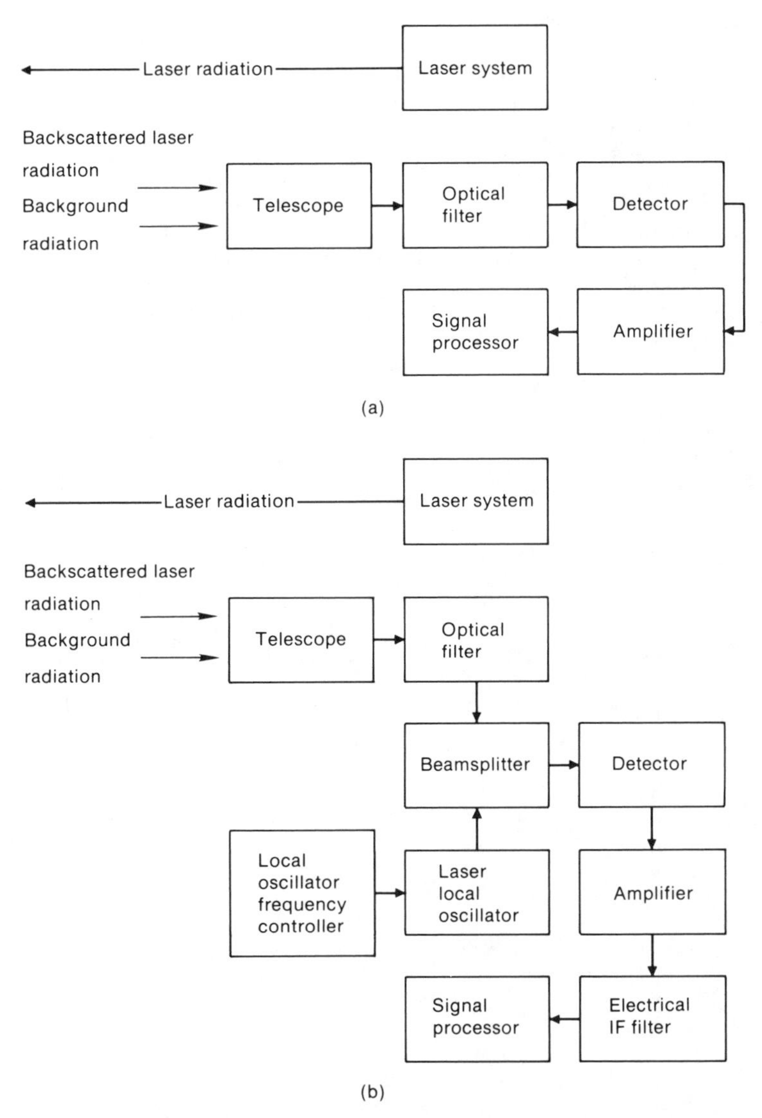

Fig. 7.1. Lidar direct detection (a) and heterodyne detection (b) systems.

the laser is operated with pulsed flash lamp excitation at room temperature. The lifetime of these lasers is limited by the pump source, but with the Space Shuttle and Space Station replacing the laser in orbit will not be a difficult task. Table 7.3 shows the laser line and its applications.

Table 7.1

CO_2 Laser Line Selection

Rotational line identification	ν (cm^{-1})	Applications
P40	924.970	Cloud-top height monitoring,
P38	927.004	vertical species profiles,
P36	929.013	cloud-top winds, stratospheric
P34	930.997	aerosol composition measurements.
P32	932.956	
P30	934.890	
P28	936.800	
P26	938.684	
P24	940.544	
P22	942.380	
P20	944.190	
P18	945.976	
P16	947.738	
P14	949.476	
P12	951.189	
P10	952.877	
P8	954.541	
P6	956.181	
P4	957.797	
P2	959.388	
R0	961.729	
R2	963.260	
R4	964.765	
R6	966.247	
R8	967.704	
R10	969.136	
R12	970.544	
R14	971.927	
R16	973.285	
R18	974.618	
R20	975.927	
R22	977.210	
R24	978.468	
R26	979.701	
R28	980.909	
R30	982.091	
R32	983.248	
R34	984.375	
R36	985.484	
R38	986.563	
R40	987.616	

Table 7.2

Nd : YAG Laser Line Selection

Wavelength (μm)	Applications
0.964	Cirrus cloud ice–water discrimination, surface albedo measurements, stratospheric aerosol profiles, ozone concentration profiles, atmospheric species vertical profiles
1.054	
1.061	
1.064	
1.074	
1.112	
1.116	
1.123	
1.318	
1.338	
1.358	

7.2.4 *Dye Laser Spectral Line Selection*

The most useful dye laser for space remote sensing applications would be one excited by a frequency-doubled Nd : YAG laser. This would have the advantages of using a space-qualified Nd : YAG laser, avoiding the degradation problem of dyes in lamp-pumped systems, and providing shorter output pulses for better range resolution. Table 7.4 shows the dye laser spectral line selection and applications.

7.2.5 *Excimer Laser Spectral Line Selection*

The excimer laser is a pulsed laser with greater than 1% electrical efficiency. Excimer lasers today cover some of the 200–400-nm spectral

Table 7.3

Alexandrite Laser Spectral Selection

Wavelength (nm)	Applications
700–800	Cloud height determination, vertical pressure and cloud-top pressure determination, lower atmospheric vertical temperature profiles

Table 7.4

Dye Laser Spectral Line Selection

Wavelength (nm)	Applications
300–1000	Cloud-top height determination, cloud ice and water discrimination, stratospheric aerosol profile sounding, surface albedo measurements, atmospheric species measurements

region. Nearly 100% coverage of this spectral region is expected in the 1990s. The laser lifetime (10^7 shots) is also of concern for space remote sensing applications, but with the Shuttle and Space Station laser replacement could be done in orbit for long-term measurements. Table 7.5 shows the excimer laser spectral line selection and applications.

7.3 Lidar System Laser Power Requirements

Remote sensing of aerosols, atomic species, molecular species, and natural surfaces from the Space Shuttle and Space Station will require some new space systems. Lidar is one of the systems that can be used for these applications. Remote sensing of atomic species usually limits the laser source to the visible or UV spectral region, while molecular species have many of their spectral bands in the infrared. Space remote sensing of clouds and aerosols is best done with lasers that operate in the visible to near-infrared regions, where scattering coefficients are large.

Table 7.5

Excimer Laser Line Selection

Wavelength (nm)	Applications
200–400	Atmospheric species concentration measurements, aerosol profile sounding

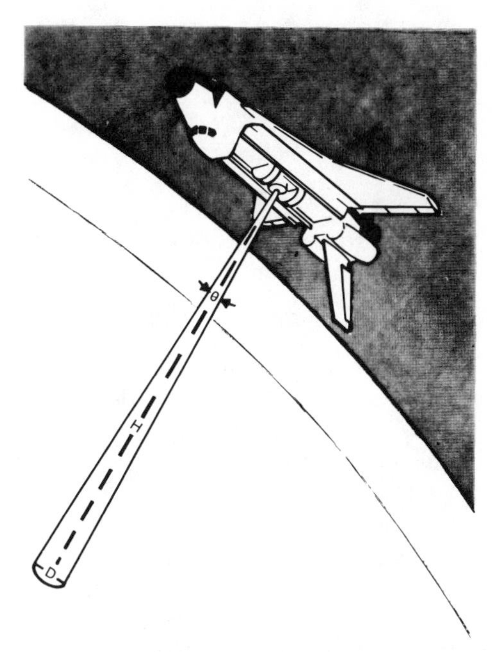

Fig. 7.2. Shuttle lidar and sounding area relationship.

Lidar remote sensing could be done from a free-flyer satellite, the Space Shuttle, or the Space Station pointing down at the earth. The laser radiation would be directed downward through the atmosphere to the earth's surface, where it would be backscattered. The backscattered laser radiation would traverse the atmosphere and be measured by the space sensor receiver on board the space system. Figure 7.2 illustrates the Shuttle lidar–sounding area relationship.

The laser power requirements for a spaceborne lidar system can be calculated as shown in the following equations. If the laser beam is directed downward and has uniform intensity, then the irradiance E at the earth's surface is given by

$$E = P/\pi R^2 \tag{7.1}$$

where P is the laser output power and R the beam radius at the surface. Considering the earth's surface as a diffuse reflector with albedo a, the reflected power I is given by

$$I = Ea/\pi = Pa/\pi^2 R^2 \tag{7.2}$$

If the lidar receiver with aperture area A_0 at height H above the earth's surface has a field of view (FOV) matching the laser beam spot on the ground the returned laser signal S at the aperture of the lidar is

$$S = IA_0\pi R^2/H^2 \tag{7.3}$$

$$= PaA_0/\pi H^2 \tag{7.4}$$

If the atmospheric transmittance is τ_A for the double atmospheric path and the optical transmittance is τ_0, then

$$S = PaA_0\tau_A\tau_0/\pi H^2 \tag{7.5}$$

Lidar systems developed to date fall into four generic categories: atmospheric window lidar, absorption lidar, differential absorption lidar (DIAL), and Doppler lidar. Absorption lidar and DIAL sensor systems can be used for atmospheric species monitoring. Doppler lidar can be used to measure wind field. Window lidar systems can be used to study the roughness of the earth's surface or aerosol vertical profiles.

Absorption lidar systems use a pulsed laser as the transmitter and a telescope as the receiver. Laser pulses are transmitted into the atmosphere, where they are backscattered by aerosols and molecules into the optical receiver. The laser output wavelength is selected to ensure that it overlaps the wavelength required for absorption by the gas molecule of interest. The backscattered radiation is detected by the lidar receiver and species abundance can be retrieved.

When the DIAL technique is used, a second laser wavelength is selected to have a minimum of absorption by the gas being sensed. The sampling volume is the same for the first and second wavelengths so that the atmospheric extinction and volume backscatter coefficients will be the same for both wavelengths. From the ratio of the signals at the two wavelengths, the gas concentration over a known range can be determined.

Doppler lidar systems are used to detect tropospheric winds by measuring the Doppler shift of the laser signal backscattered from atmospheric aerosols. The systems are capable of detecting wind speeds up to 1 m/sec. These lidar systems contain heterodyne radiometers for frequency change detection.

Atmospheric window lidar systems can be used to measure the earth's surface roughness to determine the albedo of the surface as a function of surface condition. Also, aerosol vertical profiles can be determined with this type of system.

Table 7.6 lists the laser power values for several useful spaceborne laser sources.

Table 7.6

Space Laser Power Parameters

Laser	Laser power (W)	
	1980s	1990s
CO_2	10^7	10^8
Nd : YAG	10^7	10^9
Dye	10^6	10^8
Alexandrite	10^7	10^9
HF, DF	10^7	10^8

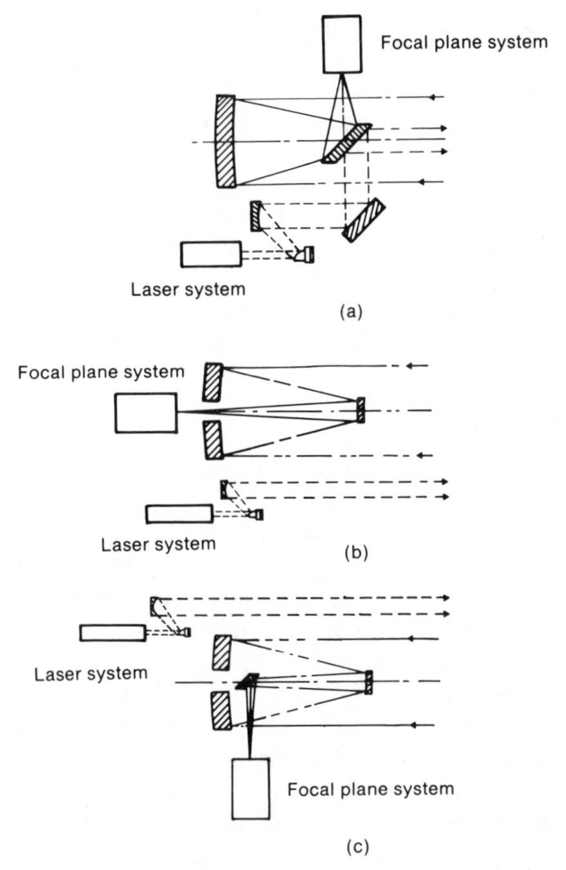

Fig. 7.3. Lidar optical system schematics: (a) Newtonian telescope; (b) Cassegrain telescope; (c) Nasmyth telescope.

7.4 Space Lidar Optical Systems

In general, lidar telescope design should match the receiver FOV to the transmitted laser pulse divergence as closely as possible, thus reducing the background brightness noise to a minimum. A reflective telescope is a better choice than a refractive telescope because of its mass and optical transmittance. A Cassegrain-type telescope, either classical Cassegrain or Dall–Kirkham, can be used as the space lidar optical transmitter and receiver for atmospheric and surface remote sensing applications. The Cassegrain type is chosen for a number of reasons, including relative ease and insensitivity of alignment, simplicity of fabrication, ease of baffling, and good throughput. Figure 7.3 shows lidar optical systems designed for space applications.

A lidar receiving telescope should have a broad spectral range of operation from the ultraviolet to the infrared region. Aluminum coating offers good reflection over the entire range of most laser sources. Other advantages of an aluminum coating are its good adherence to glass and its ability to be overcoated easily with the coating material. Thus one lidar telescope should cover the entire spectral range for space remote sensing applications.

The lidar focal plane optics can be refractive or reflective, with separate lenses or mirrors being provided for different wavelength coverage. Multichannel coverage can be achieved through rotation of the fold mirror or the use of beamsplitters for different radiometer or spectrometer measurements.

7.5 Lidar Focal Plane Detector Systems

Photomultipliers have been the most useful type of lidar detector for remote sensing in the UV and visible regions because of their fast rise time, high quantum efficiency, and high internal gain. These advantages eliminate the need for a low-noise preamplifier and permit detection of weak atmospheric radiation signals. Photomultipliers require a high-voltage power supply, and in certain weak signal applications they require cooling to reduce the dark current to acceptable levels. Many PMTs have already been used in space with very satisfactory long-life performance.

Photodiodes are also used to detect visible radiation for space applications. In the infrared region thermal or quantum detectors are used to detect atmospheric radiation or earth surface radiation for space lidar applications.

Table 7.7 lists lidar detector systems.

Table 7.7

Lidar Detector Systems

Laser	Wavelength range (μm)	Detector type	Operating temperature (K)
CO_2	9–11	HgCdTe	70–120
Nd : YAG	1.06	InGaAsP	250
Dye	0.3–1.1	PMT, InGaAsP	250–273
Alexandrite	0.7–0.8	PMT	275
HF	2.7–3.3	InSb	77
DF	3.5–3.9	InSb	77

7.6 Lidar Electrical Power Requirements

High-power laser systems require large amounts of spacecraft electrical power to support the high-energy laser generation. Typical spacecraft electrical power is near 500–1000 W. Space Shuttle electrical power for a typical 7-day mission with three fuel cells is 7000 W. The Space Station will provide 75–125 kW of electrical power for all the systems in space operation from its solar array. Table 7.8 shows the electrical power requirements of lidar systems.

7.7 Space Lidar Sensors and Applications

7.7.1 *Multiuser Shuttle Lidar Facility*

The multiuser Shuttle lidar facility (MSLF) provides unique and basic space remote sensing of the earth's atmosphere and surfaces. The Space

Table 7.8

Lidar Electrical Power Requirements

Lidar system	Electrical power requirements (W)
CO_2	4000–8000
Nd : YAG	2000–4000
Dye	1000–8000
Alexandrite	3000–6000

Station and Space Shuttle will be ideal platforms for space lidar applications. The primary goals of the lidar program are to test active lidar systems and to measure atmospheric species, clouds, and natural surfaces for studies of the earth's climate and environment and weather observations. With the lidar technique, the vertical distribution of a species is measured from observations of the magnitude of the laser backscattered radiation as a function of the time interval from laser pulsing. A typical laser pulse length of 30 nsec corresponds to a lidar range resolution of less than 5 m. The horizontal resolution of a 1-mrad laser beam at a spacecraft height of 300 km is only 300 m for high spatial sounding applications.

Space lidar systems can make an important contribution to the following objectives identified by the MSLF users' group.

(1) Cloud-top heights
(2) Tropospheric cloud and aerosol profiles
(3) Cirrus ice–water discrimination
(4) Noctilucent clouds and circumpolar particulate layer profiles
(5) Surface albedo
(6) Stratospheric aerosol profiles
(7) Alkali atom density profiles (Na, K, Li)
(8) Ionospheric metal ion distributions (Mg^+, Fe^+, Ca^+)
(9) Water-vapor profiles
(10) Trace species measurements (O_3, H_2O, NH_3, CFMs, etc.): total burden; rough profiles
(11) Chemical release diagnosis
(12) Stratospheric ozone profiles
(13) Upper-atmosphere trace species profiles
(14) Na temperature and winds
(15) Surface and cloud-top pressure measurements
(16) Tropospheric pressure profiles
(17) Tropospheric temperature profiles
(18) Trace species (O_3, H_2O, NH_3, C_2H_4, etc.) profiles
(19) Cloud-top winds
(20) Aerosol winds
(21) OH density profiles
(22) Metal atom/ion/oxide profile ($Mg/Mg^+/MgO$, 80–600 km)
(23) Tropospheric NO_2 burden profile
(24) Stratospheric aerosol composition
(25) NO density profiles (70–150 km)
(26) Atomic oxygen profiles (80–150 km)

The lidar modular source system consists of four primary subsystems: a neodymium laser and its associated modules, and dye, pulsed CO_2, and

Table 7.9

Multiuse Lidar Systems

Telescope type	Classical Cassegrain
Telescope size	1.2-m diameter
FOV	0.05–5 mrad
Coating	Al + MgF_2
Laser type	Nd : YAG, dye, pulsed CO_2, cw CO_2
Wavelength	0.532, 0.355, 0.215–0.94, 10.6, 9–11 μm
Detectors	PMTs, HgCdTe (heterodyne detection)
Total peak power	4.23-kW electrical
Total mass	1990 kg

cw CO_2 lasers and their associated modules. The neodymium and CO_2 lasers were chosen as the basic source lasers because of their demonstrated outstanding reliability, advanced state of engineering development, efficiency, tolerance to environment, compact size and weight, lack of corrosive or limited-shelf-life components, and relative simplicity. The dye laser presents a low risk and its spectral control and tuning are well understood. An alexandrite laser will be added to the source system when it is ready for space applications.

The lidar receiver system consists of telescope, lenses, filters, and detectors. Table 7.9 lists the lidar system parameters.

This multiuser Shuttle lidar facility would fly on Space Shuttle or Space Station missions. It would be mounted on a pallet and would fly on multidisciplinary missions several times a year or more. Figure 7.4 shows the lidar system arrangement.

7.7.2 *Space Doppler Lidar System*

The determination of atmospheric winds from a space platform has been a goal of many users in industry, government, and universities. Wind data are key inputs in environmental analysis and weather forecasts. Interest in space remote sensing of wind fields is very strong among users. Studies were carried out to examine the possibility of mounting lidar on the Space Shuttle and free-flyers to perform wind soundings as the beginning of a global wind-measuring satellite system called the Windsat Program. The results of the studies were very encouraging. The goal of Windsat is an operational program with a long-life space system in low earth orbit (LEO) to provide a complete observation of the earth approxi-

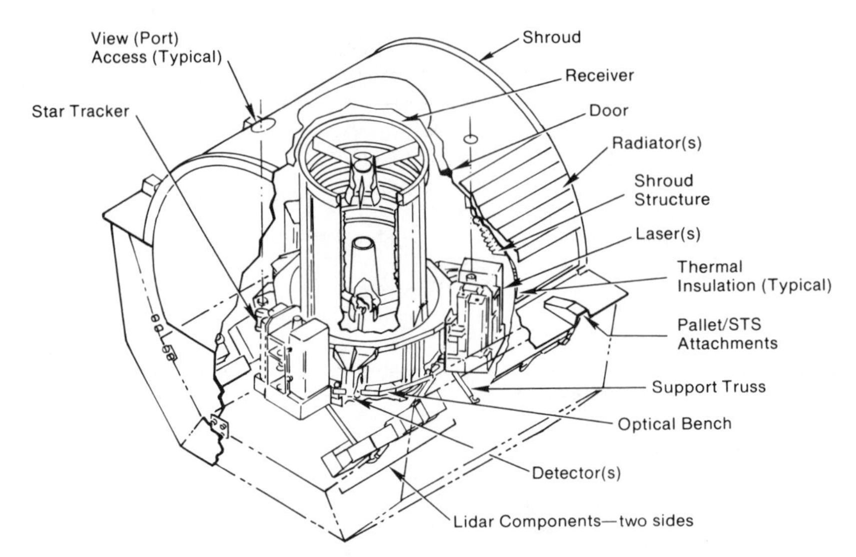

Fig. 7.4. Lidar system arrangement. (Courtesy of GE/NASA.)

mately once a day. It seems that the Space Station or other satellite system would be a good platform for wind-field sounding from space.

The Doppler lidar can be operated in scan mode or stare-scan mode. Normally, it will continuously scan at a certain number of seconds per revolution, transmitting several pulses per second and processing the backscattered signals from each of the range-gated altitudes of interest. The scan mode will provide global coverage. The stare-scan mode can be chosen by ground command for small regions of interest.

The lidar telescope is a Cassegrain-type system with a 1.25-diameter. Beryllium is chosen as the basic material for the primary mirror. The radiation source for the lidar is a pulsed, coherent, very stable, single-frequency CO_2 laser. The laser beam is directed optically to the point of interest in the atmosphere. Aerosols scatter some of the transmitted radiation in all directions and the aerosols move with the wind, which causes laser frequency shifts. From the Doppler shift of the frequency of the transmitted beam, the wind speed can be determined. If vertical velocity can be ignored, then two measurements of each volume, from different directions, are required.

The Doppler shift f_D is given by

$$f_D = 2V/\lambda \tag{7.8}$$

where V is the radial wind speed and λ the laser wavelength. For a measurement range of ± 100 m/sec, the corresponding Doppler shift will be ±21.95 MHz. To achieve a resolution of 1 m/sec, the Doppler shift must be known with a precision of 219 kHz. To differentiate between positive and negative velocities, an intermediate frequency is chosen so that zero velocity will correspond to a nonzero frequency. In addition to the lidar Doppler shift due to the aerosol winds, a larger Doppler shift associated with both the satellite motion and the earth's rotational velocity is measured.

The Doppler shift associated with the spacecraft velocity is due to the projection of the spacecraft velocity vector along the line of sight of the lidar telescope. The Doppler term can be written as

$$f_S = (2V_S/\lambda) \sin \alpha \cos \theta \tag{7.9}$$

where α is the nadir angle, θ the azimuth angle of the telescope with respect to the spacecraft velocity vector, and V_S the spacecraft velocity.

The earth's rotation will also cause a Doppler shift of the return lidar signal due to the tangential velocity of the earth. At the equator the velocity is 464 m/sec, while at the poles, it is zero. A first-order approximation to the Doppler shift caused by the earth's rotation is given by

$$f_E = (2V_E/\lambda) \sin \alpha \sin \theta \cos r \tag{7.10}$$

where V_E is the maximum earth tangential velocity (464 m/sec) and r is the lidar measurement location latitude.

Table 7.10 lists Doppler lidar system parameters for different spaceflight systems.

Table 7.10

Doppler Lidar System Parameters

Parameter	Free Flyer Lidar	Space Shuttle Lidar	Space Station Lidar
Altitude, km	830	300	300
Scan angle, deg	52.7	62	62
Ground track radius, km	1261	622	622
Vertical height & resolution, km	20-1	20-1	20-1
Slant range, km	1575	703	703
Laser energy, J	10	10	10
PRF, pps	2	8	24
Wavelength, μm	9.11	9.11, 11.99	9.11, 11.99
Laser lifetime, pulses	3×10^8	4×10^6	10^9
Heterodyne detector	HgCdTe	HgCdTe	HgCdTe

In Doppler lidar operation in space, a laser pulse is directed through a telescope toward the earth's atmosphere. The backscattered Doppler signal from the aerosols is received by the lidar receiver. The lidar telescope continuously rotates about nadir; thus the laser beam is scanned through two radius patch widths on each side of the space system ground track. Each laser pulse is overlapped in a local area with another pulse on a slightly later scan as the space system continues its orbit. The second laser pulse illuminates the same area, but at a different look angle, allowing the radial wind measurements to be converted into orthogonal vector components. The lidar heterodyne detectors have high sensitivity and allow frequency shifts to be detected. In heterodyne detection the signal from a local oscillator (LO) laser of known frequency f_{LO} is mixed with a return signal beam of frequency $f_{\mathrm{L}} + \Delta f$, where f_{L} is the transmitted laser frequency and Δf is the Doppler shift. The two laser beams are mixed on the heterodyne detector, which converts the optical signal to an electrical signal of frequency $f_{\mathrm{LO}} - f_{\mathrm{L}} \pm \Delta f$. By choosing f_{LO} properly, the electrical signal will be in a convenient microwave domain such that Δf or the wind speed can be measured with high accuracy.

The space remote sensing coherent lidar equation expressing the detector-shot-noise-limited signal-to-noise ratio S/N in a bandwidth matched to the transmitted pulse duration is

$$S/N = \pi\eta E\beta c\tau D^2 K e^{-2\rho R}/8h\nu[R^2(1 + D^2/4r_{\mathrm{a}}^2) + (\pi D^2/4\lambda)^2(1 - R/f)^2] \tag{7.11}$$

where η is the overall detector–optical system efficiency, E the transmitted pulse energy in joules, β the atmospheric backscattering coefficient in meters^{-1} steradians^{-1}, c the speed of light, 3×10^8 m/sec, τ the pulse duration in seconds, D the telescope diameter in meters, K the beam shape compensation factor, 0.46, ρ the atmospheric absorption coefficient in meters^{-1}, R the range in meters, $h\nu$ the photon energy in joules, $r_{\mathrm{a}} = 0.069\lambda^{6/5}(RC_n^2)^{-3/5}$ the turbulence-induced transverse coherence radius in meters, λ the wavelength in meters, C_n^2 the refractive-index structure parameter in meters$^{-2/3}$, and f the focal length in meters.

The space Doppler lidar performance requirement for wind velocity measurements is ± 100 m/sec range and the accuracy for wind-field measurements is ± 1 m/sec from the surface up to a 20-km height with a vertical resolution of 1.0 km.

The lidar system aperture is 1.25 m for a scanning system and the pointing requirement is 2 μrad for short-term accuracy and 100 μrad for long-term stability. The laser electrical power requirement for 10 J per pulse with 8 Hz pulse repetition frequency (PRF) is about 1800 W. Figure 7.5 shows the lidar system schematic.

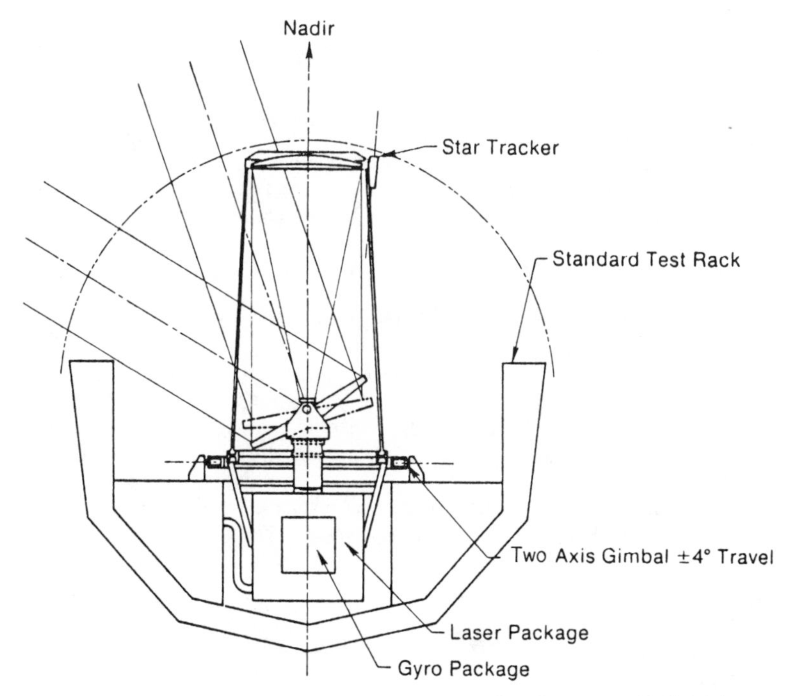

Fig. 7.5. Space Doppler lidar system schematic. (Courtesy of NOAA.)

7.7.3 *Synthetic Aperture Lidar*

Synthetic aperture lidar can be used to study the fine structure of clouds such as the mother cloud of a tornado and severe weather clouds. Lidar with a CO_2 laser having a coherent pulse duration of a few nanoseconds can be designed to study special environmental targets with a resolution of 0.1–0.5 m. Optical synthetic aperture lidar will be ready for the Space Station system in the 1990s.

General References and Bibliography

Abreau, V. J. (1980). Lidar from orbit. *Opt. Eng.* **19,** 489–493.

Browell, E. V. (1979). Shuttle atmospheric lidar research program. *NASA* [*Spec. Publ.*] *SP* **NASA SP-433.**

Greco, R. V., ed. (1980). Atmospheric lidar multi-user instrument system definition study. *NASA* [*Contract. Rep.*] *CR* **NASA-CR-**3303.

Gurk, H. M., Kaskiewicz, P. F., and Altman, W. P. (1984). Windsat free-flyer using the advanced Tiros-N satellite. *Appl. Opt.* **23,** 2537–2544.

Hinkley, E. D., ed. (1976). "Laser Monitoring of the Atmosphere." Springer-Verlag, Berlin and New York.

Huffaker, M. R., Lawrence, T. R., Madson, J. P., Priestley, J. T., Hall, F. F., Richer, R. A., and Keeler, R. J. (1984). Feasibility studies for a global wind measuring satellite system (Windsat): analysis of simulated performance. *Appl. Opt.* **23,** 2523–2536.

Killinger, D. K., and Mooradian, A., ed. (1983). "Optical and Laser Remote Sensing." Springer-Verlag, Berlin and New York.

Measures, R. M. (1984). "Laser Remote Sensing: Fundamentals and Applications." Wiley, New York.

McGuirk, M. (1981). Wind satellite (Windsat) coherent lidar pointing system. *Proc. Soc. Photo-Opt. Instrum. Eng.* **265,** 399–405.

Osmundson, J. O. (1981). Wind satellite (Windsat) experiment. *Proc. Soc. Photo-Opt. Instrum. Eng.* **265,** 395–398.

Singer, S. F. (1968). Measurement of atmospheric surface pressure with a satellite-borne laser. *Appl. Opt.* **7,** 1125–1127.

Smith, W. L., and Platt, C. M. R. (1977). A laser method of observing surface pressure and pressure–altitude and temperature profiles of the troposphere from satellites. *NOAA Tech. Memo.* **NESS-89.**

Yeh, S. D., and Browell, E. V. (1982). Shuttle lidar resonance flourescence investigations. 1: Analysis of Na and K measurements. *Appl. Opt.* **21,** 2556–2380.

Chapter 8 Active Synthetic Aperture Radar and Scatterometer Systems

8.1 Principles of Space Synthetic Aperture Radar

Synthetic aperture radar (SAR) was developed during the 1950s and 1960s as a special active microwave system to produce high-spatial-resolution surface images. The usefulness of SAR in ocean observations was demonstrated by Seasat in 1978. The Shuttle imaging radar (SIR) system has also produced good SAR research results.

In synthetic aperture imaging from space, the backscattered Doppler signal from the earth's surface is processed simultaneously with a time delay signal to generate a high-resolution image location of the surface being illuminated by the space radar beam. The imaging radar beam looks down to one side of the spacecraft and transmits many short pulses of coherent microwave energy toward the earth's surface. The strength of the returned radar signal determines the brightness of the surface. The ground resolution of an imaging radar depends on antenna size; a large antenna produces high ground resolution and a small antenna produces low resolution. It is possible to use a relatively short antenna by mounting it on a moving spacecraft and adding successive radar pulses to synthesize a long antenna.

The theoretical resolution of the SAR image, R_R or R_A, is given by the following relationships for each of the surface dimensions, where R refers to the range distance from the spacecraft to the surface. The range resolution and azimuthal resolution can be written as

$$R_R = c\tau/2 = c/2f \tag{8.1}$$

$$R_A = R\lambda/2L \tag{8.2}$$

where c is the speed of light, τ the effective radar pulse length, f the signal bandwidth, λ the wavelength, R the slant range, and L the length of the synthesized aperture.

The ground range swath width and ground azimuth width are approximately equal to

$$L_R = R\lambda/A_R \cos i \tag{8.3}$$

$$L_A = \lambda R/A_A \tag{8.4}$$

where A_A is the antenna length, A_R the antenna width, and i the incidence angle.

The synthetic aperture size L can be expressed as

$$L = R\lambda/A_A \tag{8.5}$$

Substituting Eq. (8.5) into Eq. (8.2) yields

$$R_A = R\lambda/2L = (R\lambda/2)(A_A/R\lambda) = A_A/2 \tag{8.6}$$

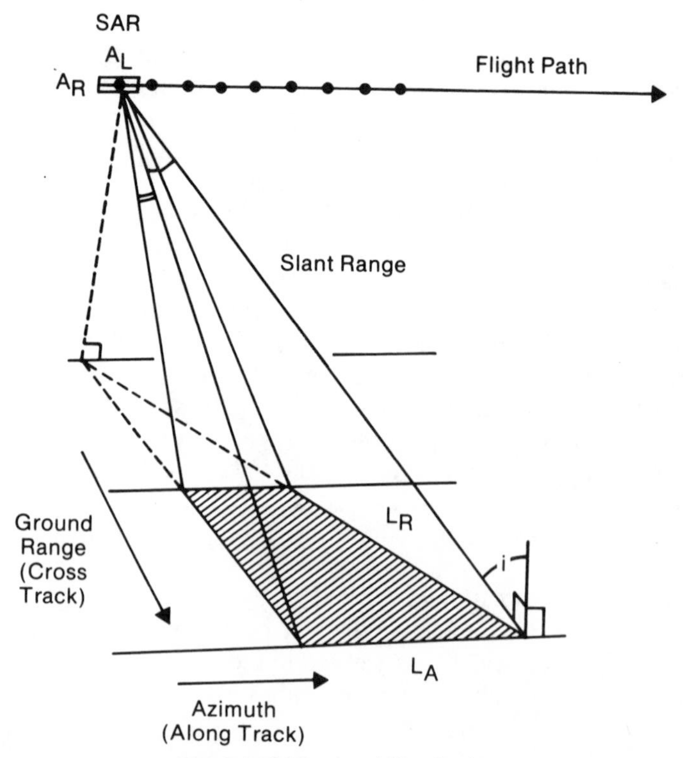

Fig. 8.1. SAR ground illumination.

Therefore azimuthal resolution is independent of the spacecraft altitude and dependent only on the antenna length.

The range and azimuth swath widths can be calculated from Eqs. (8.3) and (8.4), which resulted in 10 km and 19 km for an antenna size of 2.16 × 10.74 m (Seasat system).

The theoretical range resolution with a 19-MHz signal bandwidth is about 9 m for slant range. The ground slant range projection is 8/sin i or 18–23 m for incidence angles of 20–26°. To calculate the azimuthal resolution the antenna length is divided by 2, which yields an azimuthal resolution of about 6 m. If the illumination time or synthetic aperture is divided into four section looks, sacrificing a factor of 4 in the resolution to obtain a better signal-to-noise ratio in each ground resolution, the final resolution attained is about 25 × 25 m.

Figure 8.1 shows the SAR sensor ground illumination and resolution relative to satellite orbit.

8.2 Space SAR Spectral Frequency Selection

The spectral frequency L, S, C, and X bands are the most useful bands for space remote sensing applications. The L band covers frequencies from 0.39 to 1.55 GHz, the S band 1.55 to 4.20 GHz, the C band 4.20 to 5.75 GHz, and the X band 5.75 to 10.90 GHz. Frequency allocations are made at international conferences known as World Administrative Radio Conferences. Table 8.1 lists the allocations for remote sensing imaging applications.

8.3 Space SAR System Considerations

For space SAR multifrequency operation, reliable, powerful radar power sources are required. Solid-state sources provide good stability

Table 8.1

Imaging Radar Frequency Allocations

Band	Frequency range (GHz)	Application
L	1.215–1.300	Surface imaging
S	3.1–3.3	Surface imaging
C	5.25–5.35	Surface imaging
X	8.55–8.65	Surface imaging
	9.50–9.80	

and long operation. Calibrated modulators and antennas that will produce accurate image intensity levels for various frequencies are needed. Modular system design and construction and multimode operation are the SAR system requirements for better imaging analysis in the 1990s. The 1.274-, 5.3-, and 9.6-GHz channels are the set chosen for better space radar imaging applications.

The SAR swath width can be increased by using a stepped single beam, with a single side-looking antenna mechanically stepped or phase-shifted to achieve different selection angles, or a multibeam mode, with the beam switched in segments by multiple antennas to achieve a series of azimuthal beams. The swath width can be increased from 100 to 400 km for more ground coverage in the 1990s.

Multiple polarization and multiple incidence angle applications will be developed for space image radar systems. Horizontal–horizontal (HH), vertical–vertical (VV), and horizontal–vertical (HV) polarization models can be used for better imaging analysis. Incidence angles of 15–75° in 5° increments will supply more useful radar image information.

Table 8.2 shows the space SAR ground resolution projected to the year 2000.

Two types of SAR processors are used for data analysis: the optical processor and the digital processor. Synthetic aperture radar optical processing is a good method for turning data into images. Optical SAR images are suitable for qualitative and quick-look data analysis. The optical processor has problems such as resolution degradation, difficulty of intensity calibration, image scale variations, and geometric distortions. But optical processing of data is a fast, inexpensive method of surveying large-area data sets. It is ideal for scanning data quality and intercomparing image data. Once a specific area of interest has been identified by optical processing, the data can be digitally processed and quantitatively analyzed.

Table 8.2

SAR Operational Ground Resolution

Year	Ground resolution (m)
1978	25
1982	10
1988	5
1995	2
2000	1

Table 8.3

SAR Data Processing Time

Year	Processing time (100 × 100 km frame, 25-m resolution)
1978	8 hr/frame
1982	2 hr/frame
1984	1 hr/frame
1990s	2.5 min/frame

The digital processor converts SAR data into images that are both reproducible and suitable for quantitative analysis. Many of the problems of the optical processor are corrected by the digital processor. Although fast digital SAR processors have been built, no processor has been built that meets the practical SAR processing time requirements. The evolution of fast, economical very large scale integrated (VLSI) devices may be the solution to these requirements. Processor time is shown in Table 8.3 as a function of time.

If the spacecraft velocity is 7.5 km/sec it takes 13 sec to cover an area of 100 × 100 km, so the ratio of data processing time to data taking time is near 11 for 2.5 min/frame. This is a weak point of SAR. Another problem with SAR is the high data rate, which is 110 Mbps per channel. With a three-channel SAR system, the data rate is 330 Mbps. Thus data compression is essential for multichannel SAR, and a data compression ratio of 3 to 1 or 6 to 1 is necessary to reduce the high-data-rate problem.

The SAR sensor signal-to-noise ratio per pulse is given by the following equation:

$$S/N = PG^2\lambda^2\sigma/(4\pi)^3R^4(kT\omega)\tau \quad (8.7)$$

where P is the peak power, G the antenna gain, λ the transmitted wavelength, σ the target cross section, R the radar target range, k the Boltzmann constant, T the equivalent receiver temperature, ω the receiver noise bandwidth, and τ the system losses.

8.4 Space SAR Sensors and Applications

8.4.1 *Seasat SAR Sensor and Applications*

The Seasat synthetic aperture radar was the first imaging radar used for ocean observation. It was launched in 1978. The Seasat SAR system consisted of a planar array antenna, a transmitter and receiver sensor, an analog data link, a data formatter, a high-density digital recorder, and a processor. The Seasat antenna was a 10.74 × 2.16 m phased-array system. The antenna subsystem consisted of eight mechanically deployed, electrically coupled, flat microstrip panels. The SAR system provided the antenna with a series of high-power coherent pulses of radiation at 1.275 GHz in the L band. Linear FM chirp swept symmetrically around the center frequency with a bandwidth of 19.05 MHz and a pulse duration of 33.9 μsec. Pulse repetition rate was variable from 1463 to 1640 pps. The backscattered pulses from the surface were coupled into the receiver assembly through the output network in the transmitter. The SAR receiver output was sent to the data link along with timing and frequency references derived from the SAR system stable local oscillator (STALO), which generated a very stable signal at a nominal frequency of 91.059 MHz. The other two sections in the sensor system, the logic and control assembly and the power converter, provided the electrical interface with the spacecraft. The logic and control assembly received commands from the spacecraft, decoded them, and caused the sensor electronics subsystem to perform a certain number of operating modes. The power converter section provided the stable power required by the SAR systems. Table 8.4 lists the Seasat SAR sensor parameters, and Fig. 8.2 shows a block diagram of the Seasat SAR system.

Table 8.4

Seasat SAR System Parameters

Satellite altitude	800 km
Radar frequency	1.275 GHz
Radar wavelength	23.5 cm
System bandwidth	19 MHz
Ground resolution	25 × 25 km, 4 looks
Swath width	100 km
Antenna dimensions	10.74 × 2.16 m
Antenna looking angle	20°
Incidence angle	20–26°
Polarization	HH
Pulse repetition frequency (PRF)	1463–1640 pps
Transmitted pulse length	33.4 μsec
Transmitted peak power	1000 W
Data rate	110 Mbps

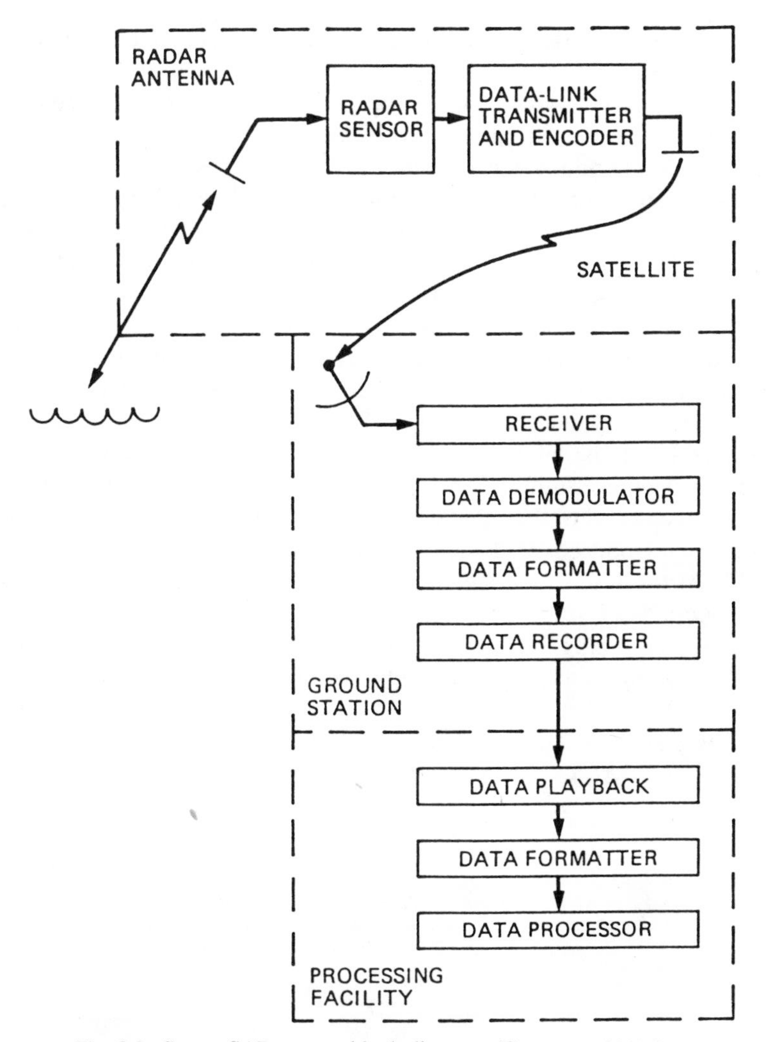

Fig. 8.2. Seasat SAR system block diagram. (Courtesy of NASA.)

8.4.2 *Shuttle Imaging Radar (SIR) System and Applications*

The Shuttle imaging radar sensor was based on the design of the Seasat SAR. Many of the subsystems were from Seasat engineering models upgraded for flight in the Space Shuttle. The SIR-A sensor antenna is a 9.4 × 2.16 m printed-circuit planar array fixed at one pointing angle. The radiating elements of the antenna are etched on copper-clad fiberglass–epoxy sheets, which are stiffened with a lightweight, honeycomb backplane. The antenna consists of seven separate panels that are summed together by using a network of coaxial cables and power combiners. The SIR-B sensor uses a movable antenna for variable-look-angle experiments. One antenna panel is added for a total of eight. The SIR-B antenna must be folded to keep its length within the confines of the pallet for launch. When it is in orbit, the antenna will be unfolded and tilted to the desired viewing geometry. During the Space Shuttle flight, the antenna can be tilted to any incidence angle between 15 and 60° from nadir.

The SIR-A optical recorder was based on a prototype left over from the Apollo 17 program. The optical recorder was modified and refurbished to provide a greater bandwidth and recording capacity than the original design. It contained 3600 ft of 70-mm film that allowed 8 hr of recording time. The SIR-B digital data-handling subsystem (DDHS) allows real-time recording on the ground of unprocessed imaging radar signal data. The SIR-B video signal to the optical recorder will be split at the output of the receiver with one channel to the optical recorder and the other to the DDHS. At the DDHS the signal is digitized with a high-speed, 6-bit analog-to-digital (A/D) converter and applied to a special memory. While one range line of imaging data is being stored in half of the memory, the other half is being dumped to the Ku-band data link of the orbiter. The Ku-band signal is relayed through a tracking and data relay satellite (TDRS) to a receiving station on the ground. The onboard optical recorder will allow recording at any time during the few days of flight.

Table 8.5

Shuttle Imaging Radar A/B Sensor Parameters

Parameters	SIR-A	SIR-B
Radar frequency, MHz	1278.4	1282
Bandwidth, MHz	6	12
Antenna size, m	9.4 × 2.16	10.7 × 2.16
Range resolution, m	40	18–50
Azimuth resolution, m	40	25
Looking angle, deg	47	15–60
Pulse width, μsec	30.4	30.4
Number of looks	6	4
Polarization	HH	HH
Data rate, Mbps	N.A.	45.5
Recorder	Optical	Optical and digital
Data processing	Optical	Optical and digital
Transmitter power, W	1100	1100

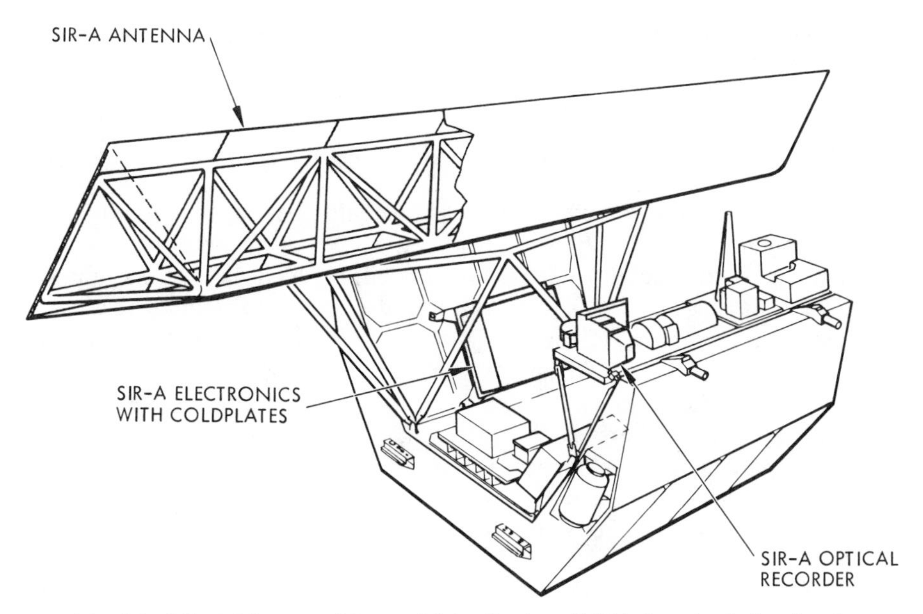

Fig. 8.3. Office of Space and Terrestrial Applications (OSTA-1) payload, showing placement of the SIR-A sensor hardware. (Courtesy of JPL and NASA.)

The SIR-A sensor and antenna were mounted on the Shuttle pallet as shown in Fig. 8.3. Table 8.5 shows the SIR-A and SIR-B sensor parameters.

8.5 Principles of the Space Scatterometer

The space scatterometer is usually used to determine surface backscattered radar energy over the ocean and, from the backscattering coefficient, derive ocean surface winds. Space scatterometers use low-resolution Doppler filtering in conjunction with real aperture techniques. This was the case with the Seasat scatterometer and will be the case with future scatterometers. The Skylab scatterometer provided a confirmation of the direct relationship between the microwave backscattering coefficient and the surface wind speed. The Seasat scatterometer provided global wind information that will be used in environmental research and weather prediction.

The backscattering coefficient is usually measured by obtaining a time average of the return signal power as the spacecraft moves. The relationship between the backscattering coefficient and the return power is shown in the following:

$$\sigma^0 = CP_{\mathrm{R}} \tag{8.8}$$

$$C = 4\pi R_c^3 \phi\beta^2 / P_{\mathrm{T}}(G/G_0)^2 L\varepsilon^2\lambda^2 L_s \tag{8.9}$$

where σ^0 is the normalized backscattering coefficient, P_{R} the radar return signal power in watts, P_{T} the peak transmitter power in watts, G the antenna gain to the center of the Doppler cell, G_0 the peak antenna gain, ϕ the antenna narrow 3-dB beamwidth in radians, β the antenna wide 3-dB beamwidth in radians, R_c the slant range to the center of the Doppler cell in kilometers, L the length of the Doppler cell along the illumination in kilometers, ε the antenna efficiency, λ the wavelength of the transmitted frequency in kilometers, L_s the system loss, and C a constant.

The space scatterometer is capable of measuring σ^0 over a range of incidence angles and determining wind velocity. The scatterometer signal-to-noise ratio can be written as

$$S/N = P_{\mathrm{R}}/kT_s B_{\mathrm{IF}} \tag{8.10}$$

where k is Boltzmann's constant, T_s the system noise temperature in degrees kelvin, and B_{IF} the IF bandwidth in hertz.

The main objective of a space multibeam microwave radar scatterometer is to measure global ocean surface winds. Surface wind data are necessary for monitoring and prediction—for example, of hazardous sea conditions—and general weather forecasting. The accuracy of surface wind measurement with the space scatterometer is ±1.6 m/sec rms in the range 3–30 m/sec. The accuracy of wind direction measurements is ±16°.

The scientific basis for scatterometer measurements from space is the Bragg scattering of microwaves from centimeter-length capillary ocean wave created by the action of the surface wind, with a backscattering coefficient which is proportional to the amplitude of the capillary waves. Also, the backscattering coefficient is anisotropic, which means that wind direction can be derived from radar measurements at two orthogonal azimuth angles. The returned radar signal data can be converted into the backscattering coefficient σ^0 and then into surface winds. Distribution of the surface wind-field data to users in a real-time manner from an operational flight system will be possible in the late 1980s or early 1990s.

Table 8.6 shows the main differences between the active microwave scatterometer and the active infrared Doppler lidar for space remote sensing of the earth's wind field. Each sensor has its special applications. It seems that the best system would be a combination of the two active sensors to overcome the weaknesses of each and provide more benefits to the user.

Table 8.6

Comparison of Scatterometer and Doppler Lidar

Parameters	Active infrared Doppler lidar	Active microwave scatterometer
Wind speed range	0–100 m/sec	3–30 m/sec
Vertical height	0–20 km	Surface only
Area coverage	Land and ocean	Ocean only
Spatial resolution	10 × 10 km	25 × 25 km
Sounding limitation	Cloud top	All weather
Wind speed accuracy	1 m/sec	1.6 m/sec
Spacecraft power requirement	1000–8000 W	110 W

8.6 Space Scatterometer Spectral Frequency Selection

The spectral frequency of the space scatterometer is usually in the K-band region from 13.4 to 14.0 GHz. Since active imaging radars use spectral frequencies in the L, C, and X bands, the best band for a scatterometer or an active sounding radar is the short end of the K band. Table 8.7 shows the spectral frequency selection for the space scatterometer.

8.7 Space Scatterometer Sensor System

Space scatterometers have been flown on Skylab and Seasat. A scatterometer is designed so that the strength of the radar backscattering coefficient is proportional to the amplitude of the surface capillary waves,

Table 8.7

Space Scatterometer Frequency Selection

Name	Spectral frequency (GHz)
Skylab scatterometer	13.9
Seasat scatterometer	14.6
NOSS[a] scatterometer	13.99

[a] National Oceanic Satellite System.

Table 8.8

Seasat Scatterometer Sensor Parameters

Type	Single-frequency fan-beam antenna
Frequency	14.6 GHz
Peak power	100 W
Pulse length	5 msec
PRF	30 Hz
Integration time	1.85 sec
Ground resolution	50 × 50 km
Antenna beamwidth	0.5 × 25°
Polarization	VV, HH
Sampling mode	(a) Both, single polarization
	(b) One side, dual polarization
	(c) One side, single polarization
Weight	102 kg

which is related to the wind speed at 10 to 20 m from the ocean surface. The scatterometer can be designed for an operational frequency of either 13.9 or 13.99 GHz. It incorporates four or six dual-polarized fan-beam antennas, which project an X-shaped pattern of illumination on the ocean

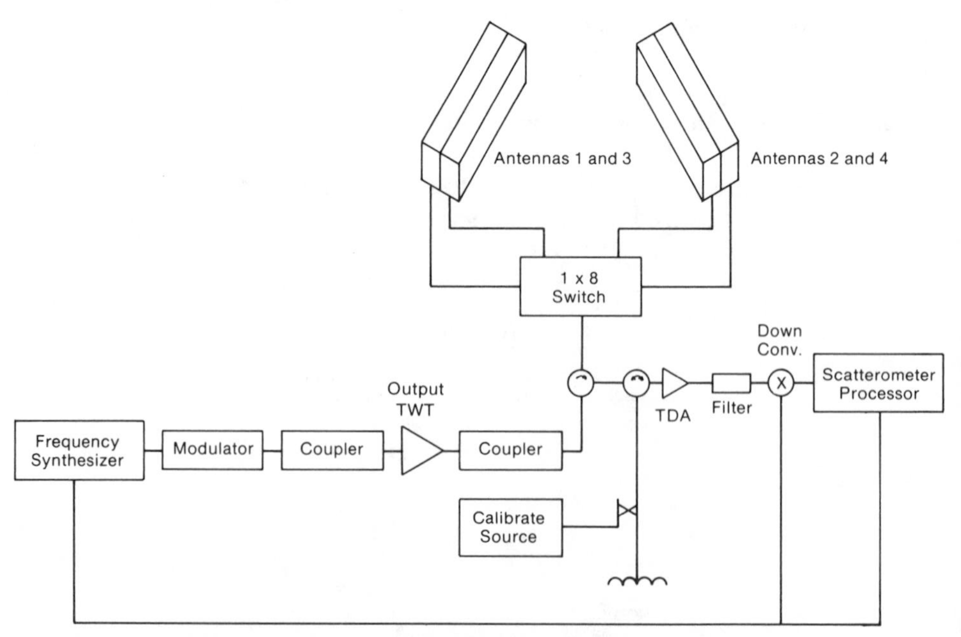

Fig. 8.4. Seasat scatterometer block diagram. (Courtesy of NASA.)

surface. For the Seasat scatterometer 12 Doppler filters were used to subdivide the antenna footprint electronically into resolution cells of approximately 50 × 50 km. The total swath covered 750 km, with the incidence angle ranging from 25 to 65° from vertical. Three additional Doppler cells provided measurements near the satellite track at incidence angles of 0, 4, and 8°.

The scatterometer sensor uses a frequency synthesizer to generate the rf signals needed by the transmitter, mixer, and scatterometer processor. The pulsed rf signal is amplified by a traveling-wave tube amplifier. The 100-W peak power output pulse is directed to the antenna, which has two parts, one for each polarization. The receiver, a square-law detector, and a gated integrator are used to sample the reflected power from an antenna 64 times during a 1.84-sec measurement period. The 1.84-sec measurement interval is repeated continuously, but a different antenna is activated for each consecutive sampling period. Table 8.8 lists the Seasat scatterometer sensor parameters. Figure 8.4 shows a block diagram of the Seasat scatterometer, and Fig. 8.5 shows the scatterometer sensor ground track characteristics.

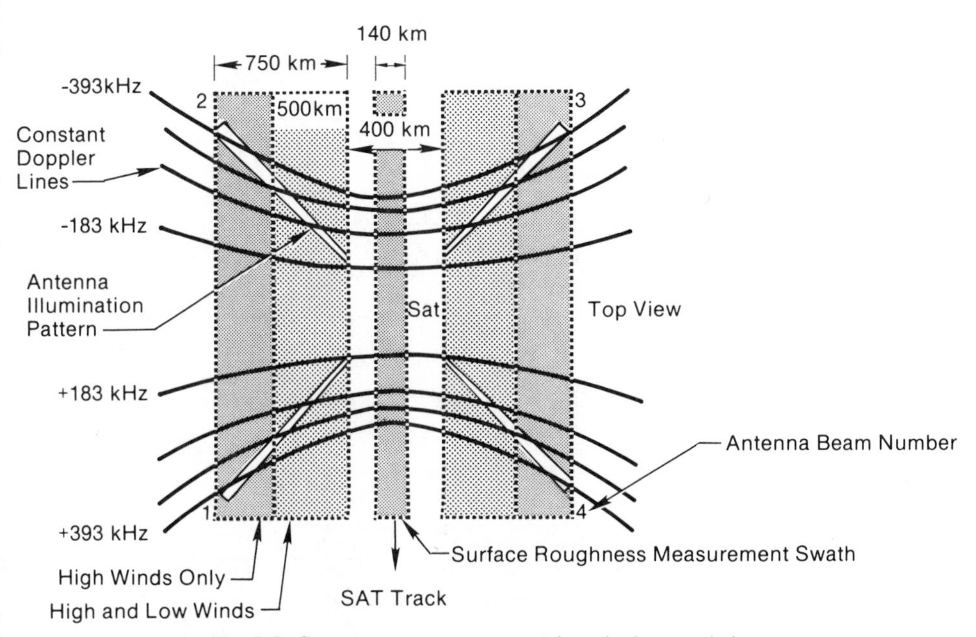

Fig. 8.5. Seasat scatterometer ground track characteristics.

8.8 Space Scatterometer Wind Measurement

A space scatterometer is sensitive to both wind direction and wind speed. The scatterometer measures the ocean radar backscattering coefficient σ^0 as a function of incidence angle. Measurements of the scattering coefficient are made in swaths at both sides of the satellite subtrack, with fan-beam antennas providing measurements in the right-hand and left-hand swaths. Either swath includes ocean surface incidence angles in the range 0 to 60°. Each antenna generates a fan-beam footprint on the surface of the ocean characterized by an azimuth angle relative to the satellite subtrack and an ocean surface incidence angle ranging from 0 to 60°. Backscattered signal energy from a single transmitted pulse is dispersed in time and frequency as determined by the geometry of the incidence angle and Doppler effects related to the motion of the satellite and the rotation of the earth. By using range gating and Doppler filtering, backscattered energy from different areas or ground resolution cells within the

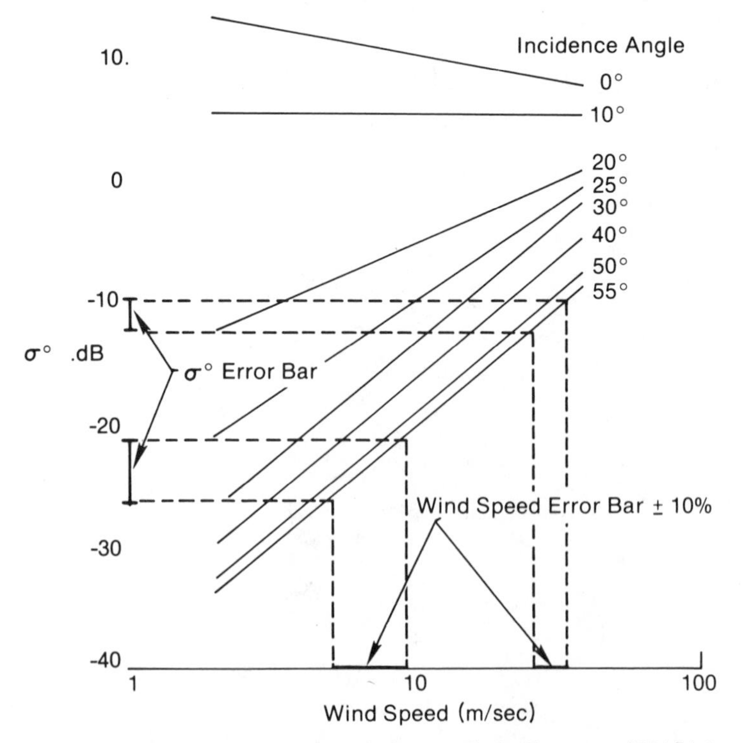

Fig. 8.6. Wind speed dependence of σ^0, crosswind. (Courtesy of NASA.)

fan beam of a single antenna is separated in the scatterometer receiver signal processor. The effective dimensions of a ground resolution cell are determined by the fan-beam antenna width, the Doppler bandwidth of the backscattered energy being measured, and the integration time of the backscattering measurement. The method of transforming wind speed accuracy requirements to backscattering coefficient σ^0 accuracy is shown in Fig. 8.6.

General References and Bibliography

Bostrom, D. E., Carnegis, G. A., Golvin, B., and Knoell, A. C. (1983). Slotted waveguide antenna for spaceborne SAR applications. Spaceborne Imaging Radar Symposium. *JPL Publ.* No. 83-11, pp. 35–40.

Carver, K. R. (1983). Spaceborne SAR sensor architecture. Spaceborne Imaging Radar Symposium. *JPL Publ.* No. 83-11, pp. 19–25.

Elachi, C. (1981). Spaceborne radar observation of the earth surface. *Pro. Int. Symp. Remote Sens. Environ., 15th,* pp. 21–32.

Elachi, C. (1983). Spaceborne radar research in the '80s. Spaceborne Imaging Radar Symposium. *JPL Publ.* No. 83-11, pp. 131–135.

Elachi, C., and Granger, J. (1982). Spaceborne imaging radars probe in depth. *IEEE Spectrum* **19**(11), 24–29.

Granger, J. L. (1983). Shuttle imaging radar-A/B sensors. Spaceborne Imaging Radar Symposium. *JPL Publ.* No. 83-11, pp. 26–31.

Grantham, W. L., Bracalente, E. M., Jones, W. L., and Johnson, W. J. (1977). The Seasat-A satellite scatterometer. *IEEE J. Oceanic Eng.* **OE-2,** 200–206.

Harger, R. O. (1970). "Synthetic Aperture Radar Systems." Academic Press, New York.

Johnson, J. W., Emedio, L. W., Bracalente, E. M., Beck, F. B., and Grantham, W. L. (1980). Seasat-A satellite scatterometer instrument evaluation. *IEEE J. Oceanic Eng.* **OE-5,** 138–144.

Jones, W. L., Schroeder, L. C., Boggs, D. H., and Wentz, F. (1982). The Seasat-A satellite scatterometer: The geophysical evaluation of remotely sensed wind vectors over the oceans. *J. Geophys. Res.* **87,** 3297–3317.

Jordan, R. L. (1980). The Seasat-A synthetic aperature radar system. *IEEE J. Oceanic Eng.* **OE-5,** 154–164.

Li, F., Winn, C., Long, D., and Geuy, C. (1984). NROSS scatterometer—an instrument for global oceanic wind observations. *Proc. Soc. Photo-Opt. Instrum. Eng.* **481,** 193–197.

Moore, R., and Fung, A. F. (1979). Radar determination of winds at sea. *Proc. IEEE* **67,** 1504–1521.

Tomiyasu, K., and Pacelli, J. L. (1983). Synthetic aperture radar imaging from an inclined geosynchronous orbit. *IEEE Trans. Geosci. Remote Sens.* **GE-21,** 324–328.

Ulaby, F. T., Moore, R. K., and Fung, A. K. (1981). "Microwave Remote Sensing: Active and Passive." Addison-Wesley, Reading, Massachusetts.

Chapter 9 | Low-Earth-Orbit Large Satellite Systems

9.1 Low Earth Orbit and Orbit Selection

With all space remote sensing systems, ground resolution, area coverage, and frequency of coverage are determined by the orbital height and inclination of the observation system. It is possible to meet some specified coverage frequency and resolution, but only by employing such a narrow field of view that incomplete surface coverage exists after the coverage interval. If a requirement of 100% coverage is imposed, then the only solution is multiple satellite systems. The space plaform altitude is an important operational parameter since it affects resolution and swath width. The higher the platform the more area that can be measured, but with a corresponding decrease in ground resolution. A larger aperture provides higher resolution, but at the cost of weight and size, so that a large launch vehicle is required. Reducing the space platform altitude produces smaller ground resolution, but at the cost of orbit perturbation due to atmospheric drag. Most LEO altitudes are in the range of 200 to 1000 km with inclination angles from 28 to 104°.

The emitted earth infrared radiation and reflected solar radiation are functions of time. There is an hourly variation of the surface temperature and reflected solar radiation. The upward earth radiation changes as a function of local time. Certain space platform requirements may dictate not only that the observations must be made at a constant local time, but also that the platform must be launched in a particular month of the year for certain area coverage. It seems that no single observation system can satisfy all operating requirements. There are two solutions to these prob-

Table 9.1

Sun-Synchronous Orbit Parameters

Orbit height (km)	Inclination (deg)	Period (min)	Orbital velocity (km/sec)
250	96.19	89.52	7.750
500	97.41	94.68	7.607
800	98.57	100.92	7.446
1000	99.49	104.10	7.344

lems. One is to increase the size of the space sensor aperture and the field of view or to use more detectors for large area coverage. The other solution is the multiple satellite systems approach for larger area coverage and greater frequency of coverage.

A wide variety of orbit inclinations and altitudes are available for space remote sensing systems. The circular sun-synchronous orbit is the most desirable for earth missions. Sun-synchronous orbits require a balance between the regression of the node and the apparent solar motion due to the earth's movement about the sun. The earth's orbital motion is about 1° per day. Sun-synchronous orbit can be achieved with the inclination angles and spacecraft altitudes shown in Table 9.1. Figure 9.1 shows the low-earth-orbital configuration.

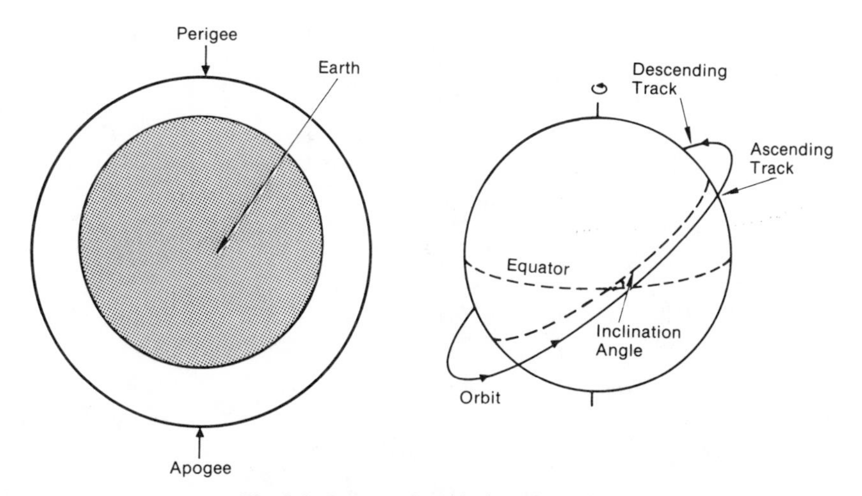

Fig. 9.1. Low-earth-orbital configuration.

9.2 LEO Unmanned Launch Vehicle

The United States has developed a number of fully integrated unmanned launch vehicle and upper stage combinations. Delta and Atlas/Centaur have been the most useful launch vehicles in the 1970s and early 1980s. The performance capabilities of those launch systems are discussed in the following sections.

9.2.1 *Delta Launch Vehicle*

Deltas are two- or three-stage vehicles consisting of a long booster with thrust augmentation supplied by multiple strap-on solid rocket motors as the first stage. The launch vehicle is approximately 1.4 m in diameter, and the third stage is a small solid rocket stage with spin stabilization as specified by mission requirements. Delta was NASA's stable launch vehicle in the 1970s and early 1980s for payloads near 1000 kg. Figure 9.2 shows the Delta vehicle configuration.

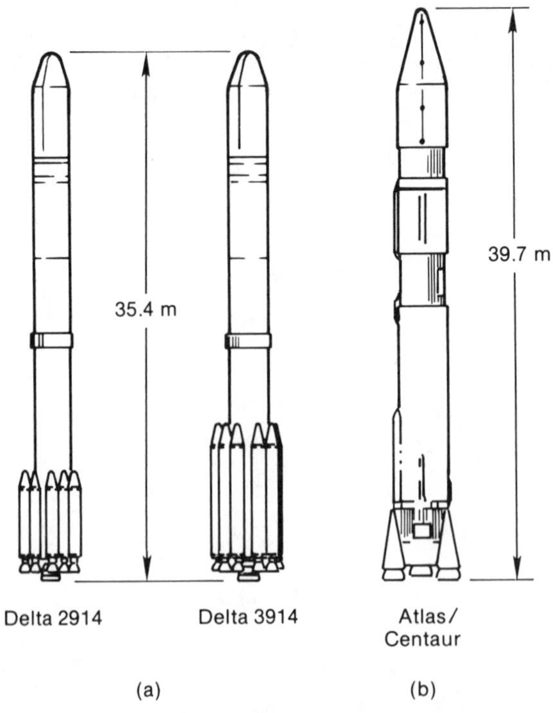

Fig. 9.2. (a) Delta and (b) Atlas/Centaur launch vehicles. (Courtesy of NASA.)

9.2.2 *Atlas/Centaur Launch Vehicle*

Atlas/Centaur was NASA's launch vehicle for unmanned missions in the 1970s. It is capable of either a direct ascent or a parking orbit flight path. The vehicle is 3 m in diameter and 39 m long and weighs 148,350 kg, excluding the payload. Usable propellant weight is 135,000 kg. From Kennedy Space Center the payload launch capability is 5000 kg for a 600-km circular orbit or 1900 kg for a geosynchronous orbit. Atlas/Centaur is a $2\frac{1}{2}$-stage launch vehicle intended as a booster for high-energy missions, particularly for geosynchronous and planetary missions. Figure 9.2 shows the Atlas/Centaur vehicle configuration.

9.3 Space Shuttle Manned Launch Vehicle

The Space Transportation System (STS) consists of the Space Shuttle, a variety of standard payload carriers, ground integration and launch facilities, and ground-based payload and operations control centers. The STS performs all the services formerly accomplished with a variety of unmanned launch vehicles. It will also routinely perform tasks not possible or practical in the past, such as manned flight operation, in-orbit servicing and maintenance, and payload retrieval and repair.

The Space Shuttle flight system consists of an orbiter, an external tank (ET), and two solid rocket boosters (SRBs). The orbiter and the solid rockets are reusable systems. The external tank is expended on each launch. The Shuttle can deliver payloads up to about 30,000 kg to a 400-km circular orbit. It can return to the earth with up to 15,000 kg of payload for reuse or repair, a capability not possible with expendable launch vehicles.

The Space Shuttle has been designed to support a wide variety of space missions, including all current and planned missions. Since it is more than a launch vehicle, it can support new missions and operations that will make possible the achievement of new space programs. The Shuttle has been designed to service and refurbish LEO satellites in orbit. On sortie missions, the Space Shuttle will transport scientists, specialists, and engineers and their experiments or observation systems into orbit for 7-day periods or longer.

The reusability of the Space Shuttle orbiter permits the construction and assembly in orbit of space station systems larger than would previously have been possible with the limitations of a single launch vehicle payload. The orbiter, in addition to its role of transport vehicle, can act as an orbiting construction base itself or provide crew support and other

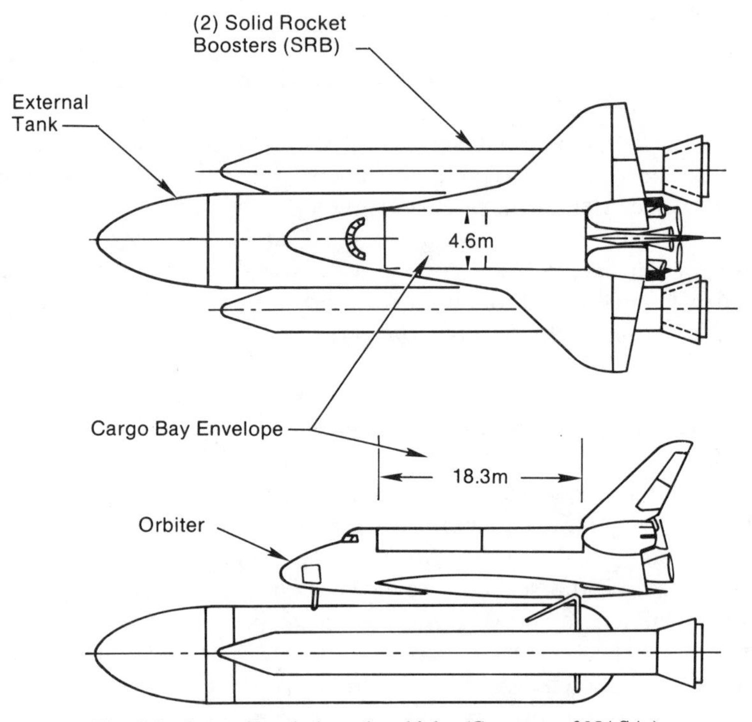

Fig. 9.3. Space Shuttle launch vehicle. (Courtesy of NASA.)

services to a space station construction facility. Figure 9.3 shows the Space Shuttle vehicle, and Table 9.2 lists the Shuttle system parameters. Figure 9.4 shows the Shuttle sensor or instrument pointing coordinate system, and Fig. 9.5 shows a typical Shuttle LEO mission profile.

9.4 Space Shuttle Pallet System

The Space Shuttle can conduct experiments in orbit with different payloads as an extension of the orbiter. It can perform most of the research and development for the earth observation experiments at a special LEO altitude. One of the most useful systems is the Shuttle pallet system, where a large number of the sensors are installed and controlled from the payload specialists' station within the orbiter. Pressure-suit operations in the payload bay are practical when instrument service is required. The Shuttle, carrying the sensor pallet into orbit, operating the

Table 9.2

Space Shuttle System Parameters[a]

Mission duration	7 days with four crewmen, 30 days with orbital maneuvering system, up to seven crewmen
Altitude, km	185–1200
Launch sites	
Kennedy Space Center	Inclination from 28.5 to 57°
Vandenberg Base	Inclination from 56.0 to 104°
Payload bay size, m	18.3 length × 4.6 diameter (maximum)
Payload weight, kg	30,000–40,000 for 28.5° inclination 15,000–20,000 for 104° inclination
Communication frequency, GHz	
Transmit	0.2968, 1.025–1.150, 1.7637–1.8398, 2.0258–2.1198, 2.205, 2.2175, 2.2875, 4.30, 13.679–13.889, 15.0034, 15.460
Receive	0.2597, 0.2790, 0.962–1.213, 1.7757, 1.8318, 2.0419, 2.1064, 2.2000–2.3009, 2.2025–2.2975, 4.30, 13.679–13.887, 13.775, 15.4–15.7
Data rate, Mbps	50, downlink
Power, KW	7 (three fuel cells)
Power supply, V dc	±27.5 to ±32.5
Pointing accuracy, mrad	8.7, orbiter internal measuring units 7.6, star sensor
Orbiter lifetime	100 flights

[a] Courtesy of NASA.

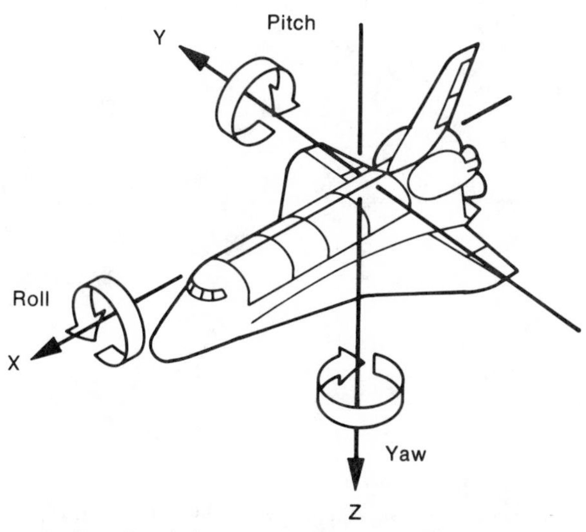

Fig. 9.4. Shuttle instrument pointing coordinate system.

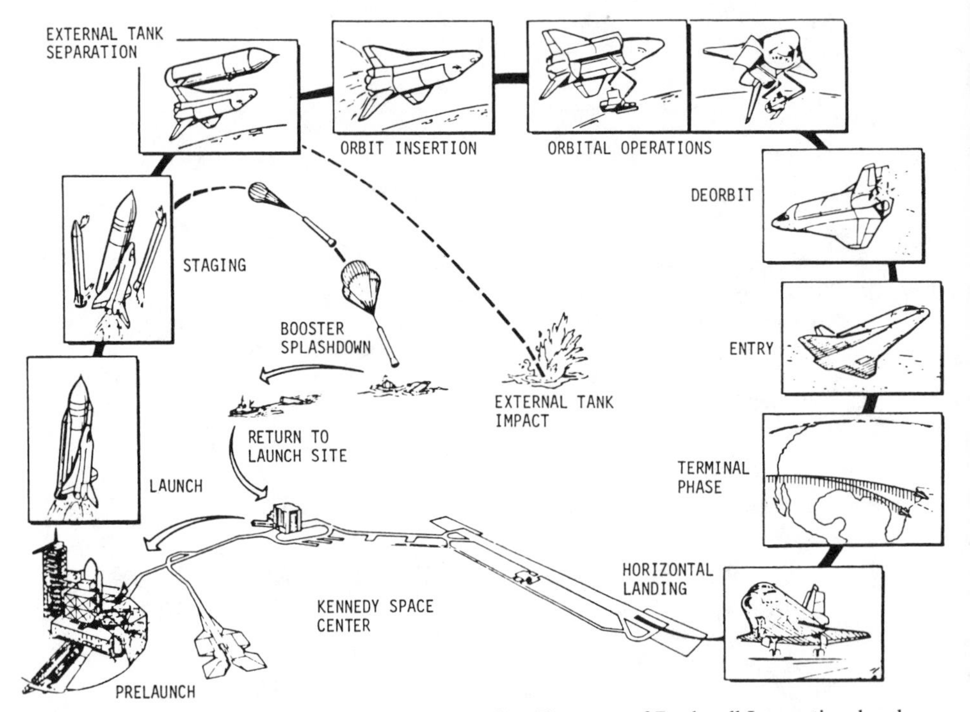

Fig. 9.5. Typical Shuttle LEO mission profile. (Courtesy of Rockwell International and NASA.)

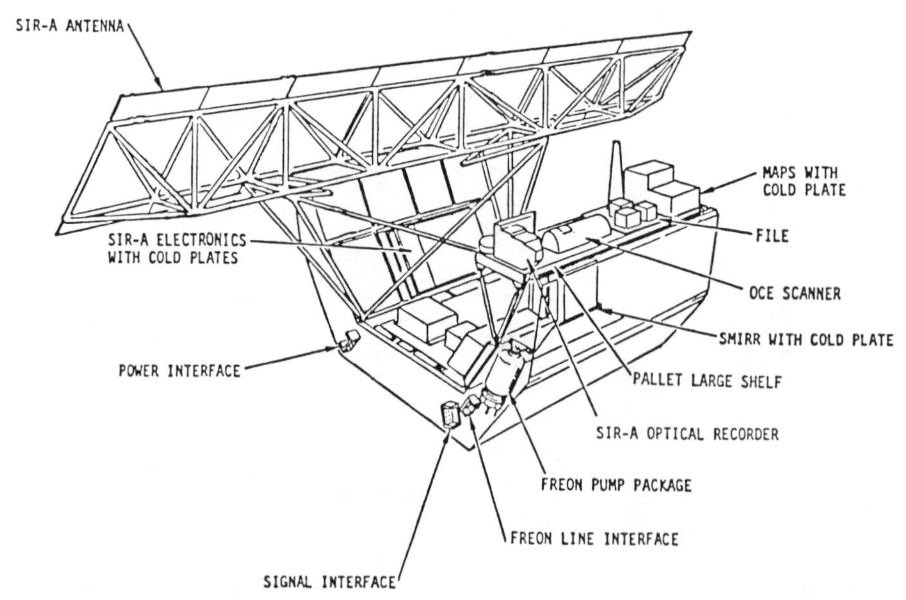

Fig. 9.6. NASA OSTA-1 configuration. See Appendix B for explanation of acronyms used. (Courtesy of NASA.)

sensors, and returning them to the earth, provides an entirely new capability for manned participation that will increase the effectiveness of space research as well as reduce the cost of science and applications.

The Space Shuttle carried the science and applications pallet payload system of the Office of Space and Terrestrial Applications (OSTA) and another payload system of the Office of Space Science (OSS) to demonstrate its research capabilities. The characteristics of the Shuttle pallet flight have been designed to accommodate the testing of flight hardware and space remote sensing concepts. During its time in orbit, the Shuttle pallet will assume an earth-viewing orientation, thus accommodating the experiments in the OSTA and OSS payloads. It is expected that experiments from many scientists and engineers will be attracted by these user-oriented features and the relatively low flight cost for reusable operation. Figure 9.6 shows the OSTA-1 configuration.

9.5 Shuttle–Spacelab System

Spacelab was developed by the European Space Research Organization for the Shuttle system. It is a reusable spaceborne laboratory that will be used to support part of the next generation of space research and applications activity. The Spacelab system can remain in orbit for 7 to 30 days, during which it acts as a short-period space station. Because of the reusability of Spacelab, operations can be repeated many times over a period of years.

The Spacelab flight vehicle consists of two basic elements: the habitable pressurized compartment module and the unpressurized equipment mounting platform pallet. The module is 4.2 m in diameter and is composed of one or two identical pressurized cylindrical shells approximately 2.7 m long, enclosed by end cones. The basic module contains a core of subsystem instruments and provides for several cubic meters of rack-installed experimental equipment. The subsystem and experiment equipment are housed in standard racks that can be removed individually during ground operations. The racks are attached to the module floor. Power and fluid lines are installed underneath. Electrical power is available for all equipment racks. Antennas, telescopes, and other sensor systems that need direct earth view or special viewing are mounted on the pallet and can be operated remotely from the module, from the Shuttle cabin, or by command link from the ground. The pallet is composed of one or many identical pallet segments approximately 3 m long, each segment being permanently equipped with standard connections to subsystem services such as electrical power. If the pallet is flown without the module, a

Fig. 9.7. Manned Space Shuttle and Spacelab overall configuration. (Courtesy of NASA.)

pressurized and thermocontrolled cylinder called an igloo provides the necessary subsystem support and interface with the Shuttle. Flight control for this configuration is provided from the Shuttle flight deck.

All the components of Spacelab were designed to a specification calling for a design life of 10 years or 50 missions of 7 days duration. Figure 9.7 shows the Spacelab–STS configuration. Table 9.3 lists the titles of the earth observation payloads on the first Spacelab mission.

9.6 Large Deployable Sensor Aperture System

The large deployable sensor system is needed for better ground resolution and large throughput in the 1990s. A rigid 5- or 10-m-diameter circular aperture will not fit into the Shuttle cargo bay. Therefore, the primary mirror must be either segmented or folded. A segmented mirror can be supported by a backup structure to form the complete mirror after assembly in orbit by the Shuttle or the space station. Development of a foldable mirror has been slow so far.

Table 9.3

Spacelab Mission 1 Earth Observation Payload

Title	Applications
Imaging UV and visible spectrometer	Atmospheric physics
Grille infrared spectrometer	and earth
Spectrometer with hydrogen and deuterium absorption cells	observations
Metric camera	
Microwave sensor	
Imaging intensifier camera	

In the development of a segmented mirror, the primary mirror segments are made from lightweight, low-expansion glass. The individual segments are supported by the backup structure at three attachment points. Each attachment point is supported by a position actuator so that the segment is adjustable in two axes. Multiple hexagonal segments, each 1–2 m across, make up the 5- or 10-m primary mirror. The whole segmented mirror can be assembled in orbit.

Deployable microwave antennas can be classified as (a) wrap-rib type, (b) radial rib type, or (c) extendable panel array type. The wrap-rib type consists of a hollow, doughnut-shaped support hub to which a series of radial ribs are attached. A lightweight mesh is supported along and stretched between these ribs to form a paraboloidal reflecting surface. The feed system is located at the prime focus of the paraboloid by three deployable support booms. Wrap-rib antennas have been built in different sizes, the largest spaceflight antenna being a 9-m-diameter aluminum-rib reflector on the ATS-6 satellite system. The radial rib antenna system consists of the ribs, feed support cone, hub, and mechanical deployment system. The rf reflective surface is formed by a dual metallic mesh configuration. The aluminum feed support cone and hub act as the primary structural backbone of the stowed antenna. Two 5-m-diameter radial rib antennas have been used in the tracking and data relay satellite system (TDRSS). A 18.3-m-diameter antenna can be made for the next generation of deployable antenna systems. The extendable panel antenna system is a linear array of pentahedral cells that provide a triangular beam cross section when fully deployed. The truss is made up of repeating panel support frames and members that connect in seven types of joints such as the Seasat synthetic aperture radar panels. The members are tubular, and all elements are made of graphite/epoxy composite with titanium end fittings. The lengths of the pivoting truss elements and the details of the joints permit both compact stowage between the accordion-folded antenna panels and member axis intersection when fully deployed.

9.7 LEO Free Flyer Systems

9.7.1 *LEO Operational Satellite System*

9.7.1.1 *TIROS-NOAA Satellite*

The first earth observation satellite, the television and infrared observation satellite (TIROS), was launched into low earth orbit in April 1960 by NASA. To provide an operational system of weather and environmental satellites, NASA decided to develop TIROS further. These TIROS operational satellites were known as the TOS series, and the satellite series was later renamed the Environmental Satellite Service Administration (ESSA) series. A second-generation operational weather and environmental satellite, the improved TIROS operational system (ITOS), was launched in January 1970. After ITOS-1 the ITOS satellites were called NOAA-1 to NOAA-5. A third-generation operational LEO polar orbiting environmental satellite system, TIROS-N, was placed in service in October 1978. The second satellite of the TIROS-N class, called NOAA-6, was launched in June 1979. The NOAA satellite series comprises the LEO operational environmental satellites of the 1980s or 1990s.

The aim of the TIROS system was to provide a series of satellites in which environmental space sensors could be tested and space-proven for the development of an operational satellite in the future. The TIROS system weighed approximately 127 kg. The stability of the TIROS spacecraft was maintained by keeping the satellite spinning about its longitudinal axis at approximately 12 rpm. The sensors carried in the TIROS satellites were basically similar and consisted of two vidicon television cameras; in approximately half of the TIROS series, a scanning infrared radiometer and an earth radiation budget sensor were added to the payload system. The TIROS orbital inclination is 48 to 58° and the orbital height 740 to 925 km.

The TOS satellite was designed as an operational environmental satellite system providing global and local imaging coverage for space remote sensing applications. It was launched into a sun-synchronous circular orbit at an altitude between 1520 and 1716 km to provide favorable illumination conditions for the cameras at all times. The orbital inclination was near 102°. The payloads were 2.54-cm tube diameter 800-line vidicons and were capable of photographing an area of 3700 km square with a 3.7-km ground resolution at the picture center. Most of the TOS satellites were launched from the Western Test Range (WTR) at Vandenberg, California. The first TOS satellite was launched from the Eastern Test Range (ETR) at Kennedy Space Center. Delta class launch vehicles were used for the TOS series.

The ITOS satellite system was launched from the Western Test Range by a Delta launch vehicle into a sun-synchronous orbit with an inclination of 101.7° and a nominal altitude of 1742 km. The ITOS satellites were designed to cross the equator at 8:35 a.m. southbound and 8:35 p.m. northbound local time each day. The evolution of the ITOS system from the proven TIROS and TOS satellites permitted growth from a spin-stabilized spacecraft to a three-axis-stabilized platform, in which each of the three orthogonal axes was held to within 0.5° throughout the mission. A momentum wheel in the pitch control system is initially spun to stabilize the satellite about the spin axis of the flywheel. Magnetic torquing coils adjust the momentum vector to the mission mode. The total weight of the ITOS satellite system is approxiamtely 340 kg. The primary space sensors aboard ITOS satellites are imagers and sounders. The imaging sensors are the scanning radiometer (SR) and the very high resolution radiometer (VHRR). The atmospheric sounder is the vertical temperature profile radiometer (VTPR). They provide surface temperatures, vertical temperature profiles, and water vapor profiles for environmental and weather applications.

The TIROS-N satellite was placed into orbit in October 1978. The second TIROS-N was named NOAA-6 for better definition or classification. The NOAA-class satellites operate in a near-polar circular sun-synchronous orbit with a nominal altitude of either 833 or 870 km and an inclination angle near 98°. In the operation configuration, two satellites are positioned with a nominal orbit plane separation of 90°. One of the satellite operates in a morning descending orbit with an equator crossing time of 7:30 a.m. and the second operates in an afternoon ascending orbit, crossing the equator at 3:00 p.m. It is a three-axis-stabilized LEO satellite system. The weight of a TIROS-N NOAA-class orbital satellite is about 736 kg (1620 lb). The sensor payload weight is about 194 kg (427 lb). The satellite electrical power system has an approximately 420-W maximum. The primary space sensors carried by the TIROS-N NOAA-class satellite are the advanced very high resolution radiometer (AVHRR) and the TIROS operational vertical sounder (TOVS), which consists of a high-resolution infrared sounder (HIRS/2), a stratospheric sounding unit, and a microwave sounding unit. The TIROS-N satellites were designed with sufficient space weight and power to allow an additional space sensor to be installed. New sensor packages currently planned for inclusion are the earth radiation budget experiment (ERBE) sensor and the solar backscatter ultraviolet radiometer (SBUV/2). The TIROS-class satellites are among the most useful LEO free-flying systems for detecting tropical storms and making weather forecasts. They are the only LEO operational satellite systems for earth environment, climate, weather, and ocean observation and monitoring from space. Figure 9.8 shows the TIROS-N NOAA-class satellite system.

9.7.1.2 *Landsat*

Landsat 1 was launched in July 1972 into a near-circular polar orbit of 903 × 921 km with an inclination angle of 99.12°. Its scientific objective was to evaluate technology involved in space remote sensing of earth resources. Landsat images are providing the basis for the mapping of detailed structure on the earth's surface. The images can be used for land survey management, estimation of water resources, prediction of agricultural products, calculation of forestry timber volume, detection of sea ice movement, and cartographic applications. Landsat 2 was launched in January 1975 into a 906 × 918 km orbit with 99.09° inclination. The Landsat orbit is sun-synchronous, which means that whenever the spacecraft passes over a given latitude the local time is the same, 9:30 AM at the equator. The swath width latitude location repeats every 18 days. Landsat 3 was launched in 1978 with the same orbit as Landsat 1 and Landsat 2. The Landsat sensors consist of a multispectral scanner (MSS) and a return beam vidicon (RBV) camera. The maximum ground resolution for the MSS is about 75 m and for the EBV is 175 m. Landsat 1 stopped functioning after more than 5 years of operation in space. Landsat 2 stopped functioning in January 1982 after nearly 7 years of space operation. Landsat 4, the second generation of the Landsat series, was launched into a 705-km circular sun-synchronous orbit at 9:30 local time in July 1982. Landsat 5 was launched in March 1984. Landsat 4 and Landsat 5 both carry two important space sensors: a multispectral scanner (MSS) and a thematic mapper (TM). The thematic mapper is a new sensor that has a ground resolution of 30 m for the visible and near-infrared bands and 120 m for the infrared window at the 10-μm band. Future TM sensor ground resolution can be improved to 10–15 m with a panchromatic wideband upgrade. Figure 9.9 illustrates the Landsat 4 satellite system. Landsat weights 1727 kg (3800 lb), has electrical power of about 2000 W, and has a data rate of about 100 Mbps.

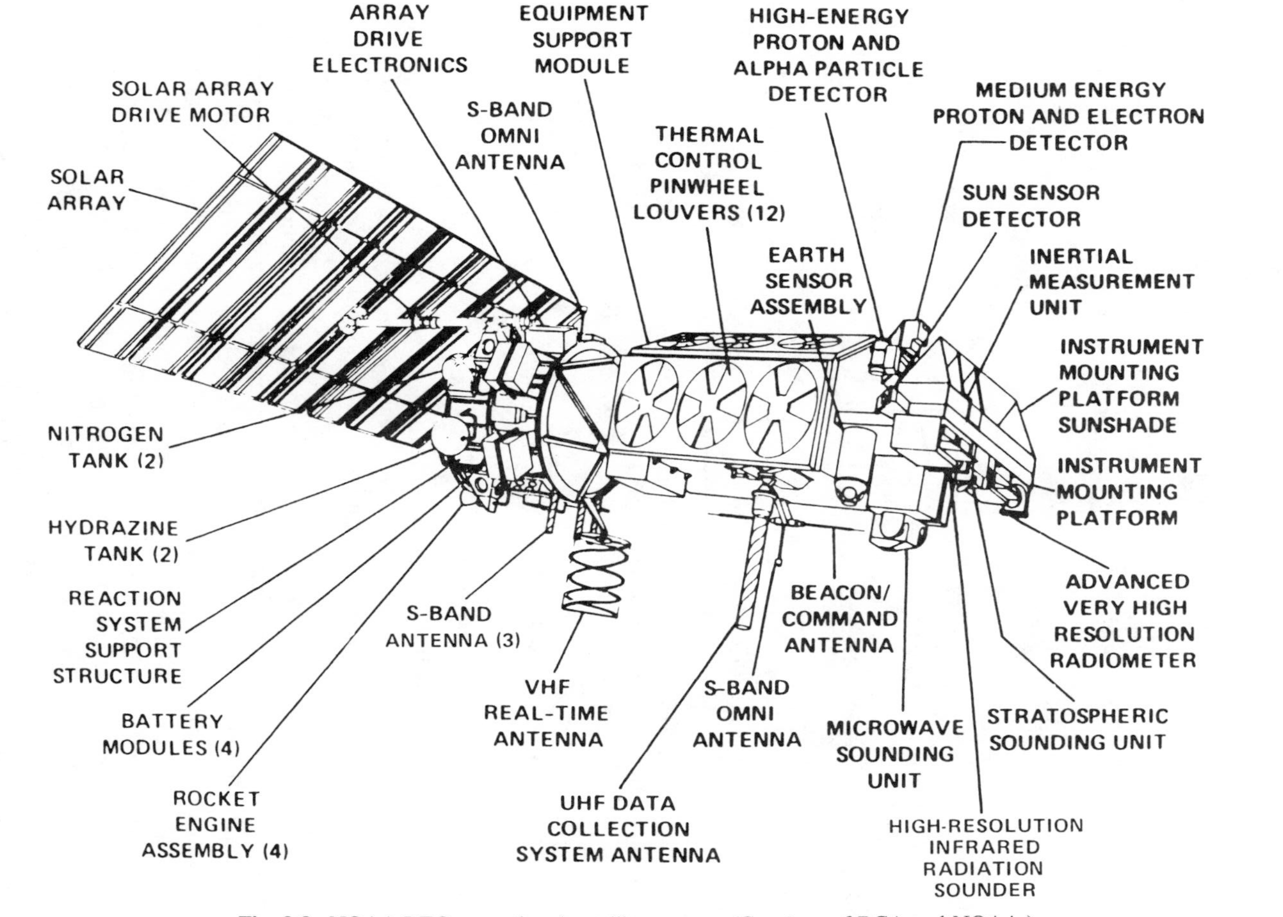

Fig. 9.8. NOAA LEO operational satellite system. (Courtesy of RCA and NOAA.)

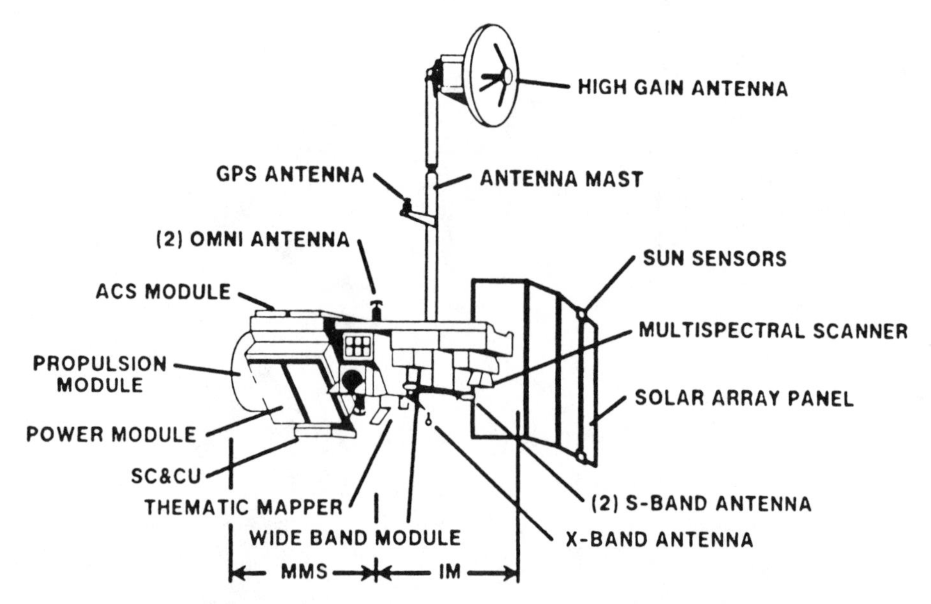

Fig. 9.9. Landsat 4 and Landsat 5 satellite system. See Appendix B for explanation of acronyms used. (Courtesy of GE and NASA.)

9.7.2 *LEO Research and Development Satellite Systems*

9.7.2.1 *Nimbus Satellites*

The Nimbus satellites were designed and developed as earth-oriented research and development satellites. They were used to measure the earth's atmosphere and surface and to provide useful research data. Nimbus was a much more advanced and expensive satellite than TIROS, partly because of the higher development cost of its advanced sensors and partly because it was a bigger three-axis-stabilized satellite. Seven Nimbus satellites have been launched into sun-synchronous orbit for different applications. Most of the Nimbus series were launched into an 1100-km sun-synchronous orbit with an inclination angle of 81° and equator crossings at noon (ascending) and midnight local time. The last Nimbus satellite, Nimbus 7, was launched into a 955-km sun-synchronous polar orbit in 1978. It carried eight space sensors: (1) a coastal zone color scanner (CZCS), (2) an earth radiation budget (ERB) sensor, (3) a limb infrared monitor of the stratosphere (LIMS), (4) a stratospheric aerosol measurement sensor (SAM II), (5) a stratospheric and mesospheric sounder (SAMS), (6) a solar backscatter ultraviolet/total ozone mapping spectrometer (SBUV/TOMS), (7) a scanning multichannel microwave radiometer (SMMR), and (8) a temperature and humidity infrared radiometer (THIR). Nimbus 6, which was launched in 1976, had seven sensors on board: (1) a temperature and humidity infrared radiometer (THIR), (2) a high-resolution infrared radiation sounder (HIRS), (3) a scanning microwave spectrometer (SCAMS), (4) an electrically scanning microwave radiometer (ESMR), (5) an earth radiation budget (ERB) sensor, (6) a limb radiance inversion radiometer (LRIR), and (7) a pressure modulator radiometer (PMR). Figure 9.10 illustrates the Nimbus 7 satellite system.

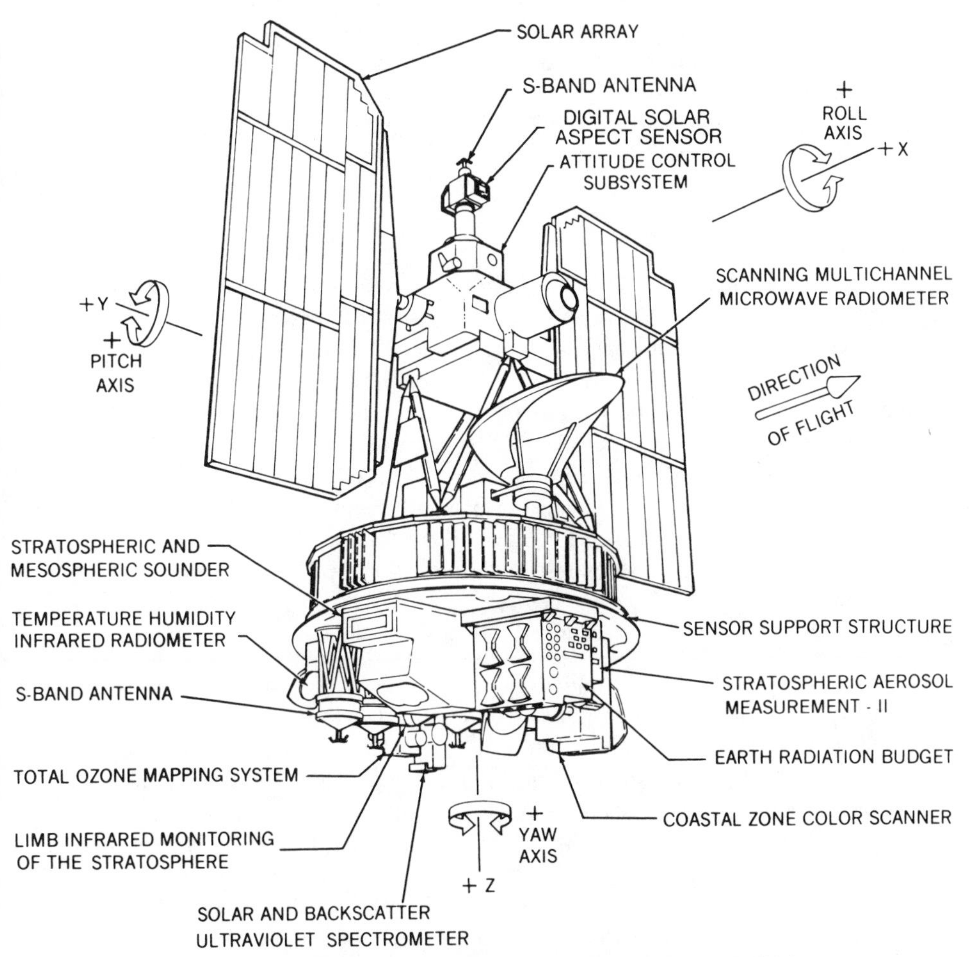

Fig. 9.10. Nimbus 7 satellite system. (Courtesy of GE and NASA.)

9.7.2.2 *Seasat*

Seasat was the first research and development satellite dedicated to the study of the global oceans with passive and active microwave sensors. Its objective was to measure sea surface wind fields and temperatures, wave heights, wave direction, currents, tides, and atmospheric water content. Seasat was launched in June 1978. There were five sensors on board the 2290-kg satellite, which had a near-circular orbit of 790-km altitude with an inclination angle of 108°. The five sensors included three active radars and two passive radiometers. The three active radars were an altimeter for precision orbital height determination, a scatterometer for ocean surface wind speed and direction measurements, and an imaging radar for surface wave images. The two passive radiometers were a microwave radiometer for ocean surface temperature and surface wind speed measurements and a visible and infrared radiometer for ocean, cloud, and surface temperature measurements. Figure 9.11 illustrates the Seasat system configuration.

9.7.2.3 *Atmosphere Explorer Mission (AEM)*

9.7.2.3.1 *Mission A: Heat Capacity Mapping Mission (HCMM).* The objectives of the HCMM were to measure day and night earth surface thermal data for rock and mineral resources location, soil moisture changes and canopy temperature, surface thermal gradients on land and in water bodies, and snow fields for water runoff prediction. The payload for the HCMM was a simple visible and infrared scanning radiometer and was launched to an orbital altitude of 620 km with a ground resolution of 0.6 × 0.6 km at nadir. The ground swath was 700 km wide, and the satellite local time was 2:30 a.m. at night followed by 1:30 p.m. in the daytime for diurnal temperature maximum and minimum measurements.

9.7.2.3.2 *Mission B: Stratospheric Aerosol and Gas Experiment (SAGE).* The SAGE satellite system was launched into orbit in February 1979 to monitor the concentrations and distributions of stratospheric aerosols, nitrogen dioxide, and other species that affect the radiation dynamics of the earth's atmosphere. The SAGE-1 satellite lasted about 3 years in space and provided especially important data on the large volcanic eruptions of Mount St. Helens. The SAGE sensor was a solar occultation sensor that looked at the limb atmosphere through sunrise and sunset to monitor atmospheric species. The orbital altitude of SAGE was near 600 km with an inclination angle of 50° for equator-to-mid-latitude coverage.

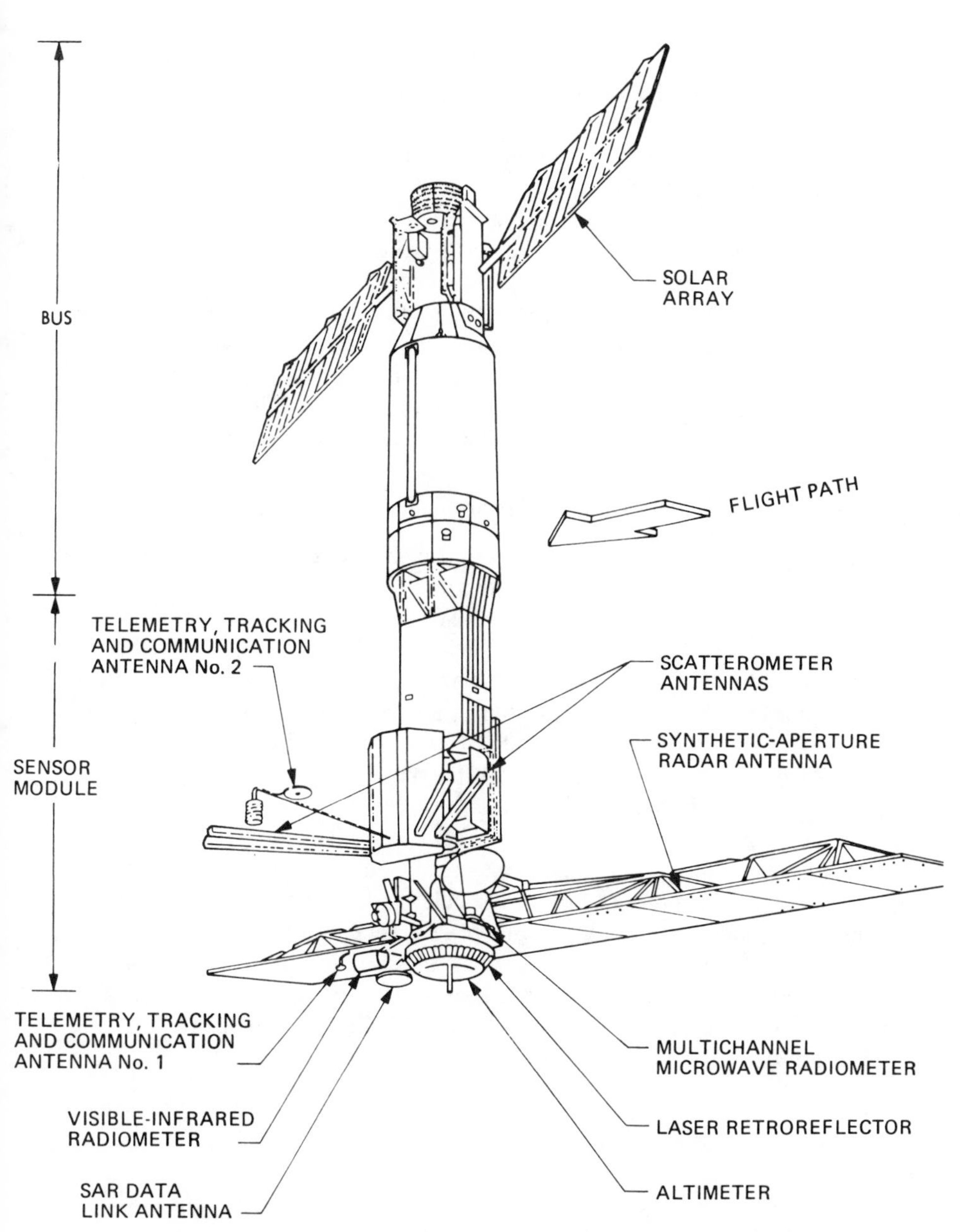

Fig. 9.11. Seasat flight system. (Courtesy of Lockheed and NASA.)

9.7.2.4 *Earth Radiation Budget Satellite (ERBS)*

The objective of ERBS is to measure the radiation equilibrium that exists in the earth–atmosphere system between the amount of solar radiation the earth takes in and the amount of thermal radiation it emits. The earth radiation budget is one of the most important factors influencing climate changes. Two other NOAA satellites also carry ERB sensors identical to that on ERBS. The three satellites participate in the earth radiation budget experiment program. The orbital altitude of ERBS is near 600 km with a 46° inclination angle so that it can monitor the middle latitudes, where sun and earth interactions are most intense. The ERBS system is to be launched by the Space Shuttle in October 1984. The payload for ERBS consists of a nonscanning sensor for limb-to-limb coverage and a scanning sensor with 46-km resolution at nadir.

9.7.2.5 *Upper Atmosphere Research Satellite (UARS)*

The objectives of the UARS system are to measure solar irradiance at ultraviolet wavelengths, to monitor the distribution and time variations of upper atmospheric species, and to measure the dynamic wind field of the upper atmosphere. With all these measurements users can study the atmospheric physical processes acting within and upon the stratosphere, mesosphere, and lower thermosphere. The UARS launch date is in the late 1980s. The satellite mass is about 3000 kg and the electrical power about 2000 W. The sensors on board the spacecraft are two visible interferometers (one Michelson and one Fabry–Perot) for temperature, wind, and species measurements, one infrared Fabry–Perot interferometer for atmospheric species monitoring, one gas filter high-resolution sounder for limb atmospheric temperature and species measurements, one gas filter sounder for solar occultation atmospheric species monitoring, two UV spectrometers for solar radiation and star radiation measurements, one microwave radiometer for limb atmospheric species and temperature measurements, and one charged-particle monitor for measurements of upper atmospheric high-energy particles.

9.7.3 *Future LEO Free-Flyer Systems*

New technology has been developed in the past several years and new space systems will be ready for launch in the 1990s. It seems that the global wind research satellite, global cloud research satellite, global environment research satellite, and ocean, land and atmosphere research satellite (platform type) will be the systems for the 1990s and beyond. For

Table 9.4

Global Wind Research Satellite

Space sensor system	Applications
Doppler lidar system	0–20-km wind field
Scatterometer system	Surface wind field
Fabry–Perot interferometer system	15–140-km wind field
Michelson interferometer system	80–150-km wind field
Gas filter sounder system	30–130-km wind field
Microwave sounder system	70–110-km wind field

example, for global wind-field measurements from space, the sensors needed are shown in Table 9.4. The global cloud research satellite sensors needed for future cloud measurements from space are listed in Table 9.5.

The ocean, land, and atmosphere research satellite will consist of the following sensors: moderate field of view (FOV) imaging radiometer, wide FOV imaging radiometer, thematic mapper, radar altimeter, ocean wave directional spectrometer, scanning multifrequency microwave radiometer, radar scatterometer, IR/microwave sounder, microwave limb sounder, UV spectrometer limb sounder, IR gas filter sounder, solar constant monitor, imaging spectrometer, high-resolution imaging radiometer, thermal IR multispectral imager, scanning laser altimeter, multifrequency lidar facility, real aperture radar, fast scanning large-aperture microwave radiometer/scatterometer, and multifrequency, multipolarization, multilook-angle synthetic aperture radar (SAR).

The new Space Station for science and applications will be launched in the 1990s. Some of the future free-flyer systems will be part of the manned Space Station payload. The next section will present the relationship between the Space Station and earth observations.

Table 9.5

Global Cloud Research Satellite

Space sensor system	Applications
Multiuser lidar facility	Cloud monitoring
Multispectral cloud radiometer	Cloud classification
Laser altimeter	Cloud height mapping
Synthetic aperture lidar	High-resolution cloud imaging
Active polarization lidar	Cloud classification
Imaging spectrometer	High spectral cloud imaging

9.8 LEO Space Station Sensor Systems

Earth observations from space include studies that utilize the wide coverage of the earth made possible by space systems to investigate the earth's climate, environment, land, ocean, and weather. The earth and its atmospheric system can be measured by pointing either an active or passive sensor either downward or toward the limb of the earth. A unique capability for earth observation will arise from the permanent presence in space of men and women and their interaction with the space instrumentation. A manned Space Station will benefit from the development of advanced technologies across many scientific and engineering fields. Space sensor systems will benefit from routine servicing in order to upgrade or repair hardware in orbit.

A Space Station at 28.5° inclination would offer the opportunity to conduct regional research across a range of earth observations, but not on a global scale. A Space Station in near-polar orbit would offer the opportunity to address problems on a global basis. Earth science and monitoring from the Space Station will be discussed in the following sections.

9.8.1 *Climate Observation from the Space Station*

In recent years the observation of climate from space has become an important area. Space sensors offer a variety of measurement methods for global quantitative observations of atmospheric and surface parameters with important climatological applications. The climate problem is one of the most critical issues facing the world today. Since climate must be observed globally, there has been great interest in space remote sensing for climate observation in recent years. The climate remote sensing probram must be regarded as one of the most important elements of future earth observation space systems.

The Space Station will be an ideal space system for the large and complex space sensors to be used for climate monitoring missions. The active space lidar system and the large passive microwave radiometer are good candidate sensor systems for part of the Space Station payload. Earth radiation budget space sensors can be used to measure the spatial and temporal components of the earth's radiation. These space remote sensing data can be used to develop climatology models needed for long-term climate prediction. Understanding the global balance between the absorption of solar radiation and the emission of the earth's radiation is considered by scientists to be pivotal to understanding the physical pro-

Table 9.6

Climate Space Sensor Requirements

Sensor type	Applications
Lidar system	Cloud physical parameters
	Atmospheric species
	Vertical temperature
Microwave radiometer	Surface temperature
	Vertical temperature
	Atmospheric species
Infrared interferometer	Atmospheric species
	Vertical temperature
Radiation budget radiometer	Earth radiation budget
(1) UV–visible–IR sensor	Radiation climatology
(2) Microwave sensor	Effect of clouds on radiation

cesses that govern the earth's climate. Climate at any time consists of long-term averages of various atmospheric variables. The radiation budget sensors can be designed for radiation measurements in the UV–visible–IR with a cloudy atmosphere and in the microwave under cloudless conditions. Table 9.6 lists the climate space sensor requirements for the Space Station system.

9.8.2 *Environment Observation from the Space Station*

Use of the Space Station for earth environmental monitoring could capitalize on the capabilities of the active lidar sensor and high-resolution spectrometer to measure atmospheric parameters. Pollutants in the earth's atmosphere cause damage to man and the environment. Because the Space Station will have the ability to overfly the entire globe, space sensors carried aboard the station can obtain overall information on this global problem. The station can also make detailed observations on a local or regional scale to measure the variations of earth and atmospheric species as a function of time.

Vertical profiles of atmospheric constituents can be measured by using a high-resolution spectrometer to obtain infrared absorption profiles by viewing the earth's limb or the sun during Space Station sunrise or sunset. Another technique is to use lidar to measure atmospheric and surface species profiles from the Space Station. Table 9.7 lists the sensor requirements for environmental observations from the Space Station.

Table 9.7

Environmental Space Sensor Requirements

Sensor type	Applications
Gas filter radiometer	Atmospheric species profiles
Active lidar	Atmospheric species profiles, surface pollution
Inteferometer spectrometer	Atmospheric species profiles

9.8.3 *Land Observation from the Space Station*

Mankind is changing the natural land cover through actions such as industrialization, deforestation, urbanization, and disposal of waste materials. The scale of human activity is clearly sufficient to alter the land environment regionally, and it is only a matter of time before it reaches a global level. These changes could result in alternations to global atmospheric parameters and land productivity. The Space Station can provide the opportunity to obtain land observation data for new surveys and detailed studies. From global land data for extended periods of time, secular and seasonal changes could be separated from long-term changes in the land cover.

Monitoring the land by space remote sensing is far more economical and practical than conventional methods involving airplanes and ground surveys. A number of potentially dangerous phenomena, including volcanism, earthquakes, and landslides, can be studied by the Space Station.

With the advent of the Space Station in the mid-1990s, users will be able to employ new visible, infrared, and microwave imagery to identify land surface structures. An advantage offered by the Space Station will be its ability to support many large space sensors so that data may be acquired over a particular area simultaneously with all sensors. Table 9.8 lists the land observation sensor requirements.

9.8.4 *Ocean Observation from the Space Station*

The Space Station will be a useful system for determining the circulation, heat content, and horizontal heat flux of the global oceans, ocean–atmosphere relationship, and variation of polar sea ice cover. The Space Station will be a good space platform for the deployment and operation of ocean sensors such as the visible ocean color imager, SAR, scatterome-

Table 9.8

Land Observation Space Sensor Requirements

Sensor type	Applications
Active lidar multispectral/ multipolarization/ multi-look-angle MLA[a]	Land imagery
Active radar multifrequency/ multipolarization/ multi-look-angle SAR	Land imagery
Passive polarization MLA	Land imagery

[a] MLA = multispectral linear array.

ter, and large-aperture microwave radiometer to measure global sea surface height, ocean surface wind velocity, ice boundaries, ocean surface temperature, and chlorophyll concentrations. In addition, the Space Station will make an excellent platform for the inspection and evaluation of new ocean sensors in orbit. Space remote sensing of the oceans could be of tremendous benefit both conceptually and as an aid to intensive investigations of specific ocean problems. Table 9.9 lists ocean sensor requirements for the Space Station.

Table 9.9

Ocean Sensor Requirements

Sensor type	Applications
Ocean color imager	Chlorophyll concentration
Synthetic aperture radar	Ocean wave spectrum and location, sea ice extent and location
Large-aperture microwave radiometer	Sea surface temperature, ice cover, atmospheric water vapor
Scatterometer	Wind speed and direction over ocean
Lidar and radar altimeter	Ocean surface topography

9.8.5 *Weather Observation from the Space Station*

Remote sensing from space has already shown great promise in providing measurements needed for studies in the atmospheric sciences on subjects ranging from severe thunderstorm lightning to the earth's weather circulation. Advanced space sensor technology appears to be capable of providing space sounding that surpasses ground measurements in quality, while retaining the global coverage only satellites can provide. The Space Station observations will ultimately constitute the most effective and economical means of acquiring the data needed to describe atmospheric parameters ranging from local to global in scale. Atmospheric temperature can be measured with space infrared and microwave sensors. Active space lidar can be used for water vapor profile measurements. Active space radar can be used for measurements of precipitation and liquid water content profiles. Clouds are principal elements of the changing weather and are vital to all life on the earth because of their role in precipitation. Space remote sensing of clouds is a major objective of the Space Station system. Lidar is one of the most useful sensors for measuring water clouds and ice clouds and their movement, location, height, and global distribution.

Atmospheric wind fields represent the dynamics of the air and are the result of atmospheric heating and cooling and of the earth's rotation. A new technique for measuring atmospheric wind fields involves the use of infrared lidar to measure the Doppler shift caused by aerosols drifting with the wind. The Space Station will be the place to test this high-power lidar system for wind observation from space. Table 9.10 list weather sensors requirement for the Space Station system.

Table 9.10

Weather Sensor Requirements

Sensor type	Applications
Active wind lidar	Atmospheric wind field
Advanced infrared sounder	High-resolution temperature profiles
Active cloud lidar	Cloud type, height, content, lifetime
Active atmospheric species lidar	Atmospheric temperature, pressure, and water vapor profiles
Active radar	Rain precipitation

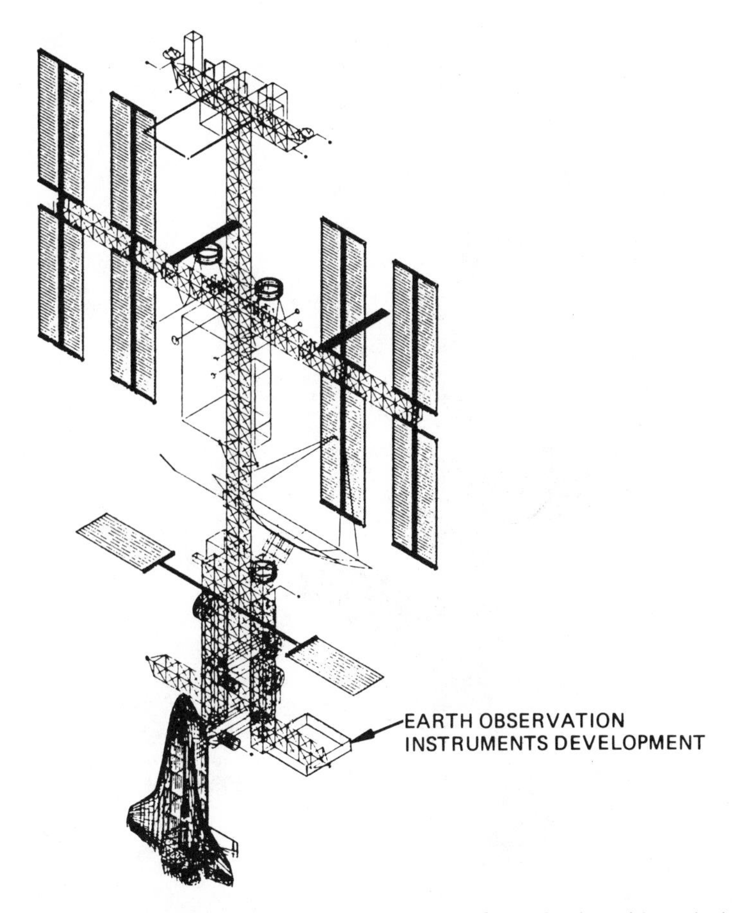

Fig. 9.12. A three-dimensional view of the LEO Earth Space Station with earth observation instruments development platform attached to it in the early 1990s. (Courtesy of NASA.)

In the design of space sensors for the Space Station system, a modular hardware approach can be adopted so that individual sensor components are readily interchangeable and can be replaced in orbit to accommodate changing measurement objectives. Periodic manned service in orbit could be used to replace aged components or reconfigure the sensor system for multidisciplinary objectives, as well as to provide routine maintenance and alignment. The frequency of such manned service would vary from daily to once a week or once a month. The goal of such sensor service would be to improve sensor performance and in the long run to reduce costs.

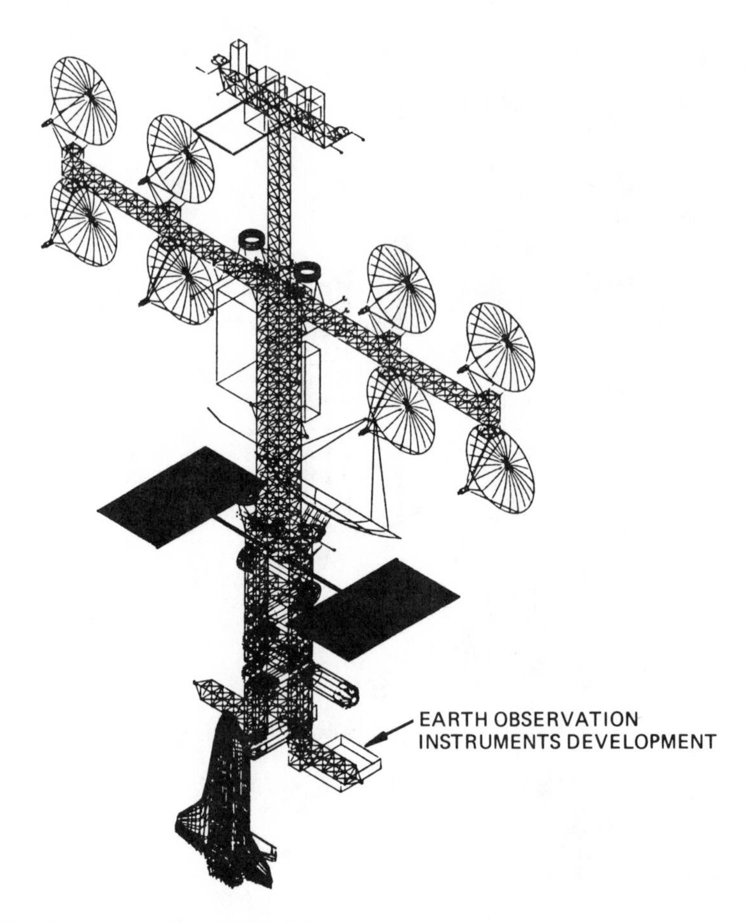

Fig. 9.13. A three-dimensional view of the growth configuration of the LEO Earth Space Station with earth observation instruments development platform attached to it in the late 1990s. (Courtesy of NASA.)

Figures 9.12 and 9.13 illustrate the initial LEO Earth Space Station in the early 1990s and the growth configuration of the LEO Earth Space Station in the late 1990s. The growth version would increase the Space Station's electrical power capability from 75 to 300 kw, and the large solar panels would be replaced by parabolic solar radiation collectors. The earth observation instruments development platform would be one of the science and application mission items attached to the Space Station.

General References and Bibliography

Allison, I. J., Wexler, R. L., and Bandeen, W. R. (1977). "Remote Sensing of the Atmosphere from Environmental Satellite," Rep. X-901-77-132. Goddard Space Flight Cent., Greenbelt, Maryland.

Atlas, D., ed. (1980). "Climate Observing System Studies: An Element of the NASA Climate Research Program Workship Report." Goddard Space Flight Cent., Greenbelt, Maryland.

Banks, P. M., ed. (1978). Upper atmosphere research satellite program. *JPL Publ.* No. 78-54.

Broome, D. R. (1983). Future remote sensing instruments and systems. *Proc. Int. Symp. Remote Sens. Environ., 17th,* pp. 65–74.

Cleven, G. C., Neilson, R. A., and Yamarone, C. A. (1983). Topex: Watershed coming in oceanography. *Astronaut. Aeronaut.* **11,** 60–65.

Hass, I. S., and Shapiro, R. (1982). The Nimbus satellite system–remote sensing R&D platform of the 70's. Proceedings of the AIAA 20th Aerospace Sciences. *NASA Conf. Publ.* No. 2227, pp. 17–29.

Hodge, J. D. (1984). The space station program plan. *Aerosp. Am.* **9,** 56–59.

Houghton, J. T. (1979). The future role of observations from meteorological satellites. *Q. J. R. Meteorol. Soc.* **105,** 1–23.

Kopia, L. P., and Luther, M. R. (1981). Earth radiation budget experiment instrument design status. *Conf. Atmos. Radiat., 4th Am. Meteorol. Soc.,* pp. 187–194.

Lame, D. B., Born, G. H., Dumme, A. J., Spear, A. J., and Yamarm, C. A. (1980). Seasat performance evaluation: the first two steps. *IEEE J. Oceanic Eng.* **OE-5,** 72–73.

Mitchell, K. L., and Wilson, G. S. (1982). Space platforms and meteorological applications. Proceedings of the AIAA 20th Aerospace Sciences. *NASA Conf. Publ.* No. 2227, pp. 43–51.

Price, J. C. (1978). Heat capacity mapping mission. *JBIS, J. Br. Interplanet. Soc.* **31,** 313–316.

Salomonson, V. V., and Koffler, R. (1983). An overview of Landsat-4 status and results. *Proc. Int. Symp. Remote Sens. Environ., 17th,* pp. 279–292.

Scherman, J. W. (1977). Current and future satellites for oceanic monitoring. *Proc. Int. Symp. Remote Sens. Environ., 11th,* pp. 279–298.

Schnapf, A. (1980). TIROS-N operational environmental satellites of the 80's. *J. Spacecr.* **18,** 172–177.

Schwalb, A. (1978). TIROS-N/NOAA A–G satellite series. *NOAA Tech. Memo.* **NESS-95.**

Staelin, D. H. (1977). Atmospheric sounding with passive microwaves: review and prognosis. *Proc. Int. Symp. Remote Sens. Environ., 11th,* pp. 401–406.

Yates, H. W., ed. (1984). Recent advances in civil remote sensing. *Proc. Soc. Photo-Opt. Instrum. Eng.* **481.**

Chapter 10 | Geosynchronous-Earth-Orbit Large Satellite Systems

10.1 Geosynchronous Earth Orbit and Orbit Selection

Most of the geosynchronous-earth-orbit (GEO) satellite systems utilize spacecraft placed in geostationary orbits. The geostationary space system is in a circular orbit on a plane identical with the earth's equatorial plane and at an altitude such that the orbital period is identical with the earth's 24-hour rotational period. This altitude is approximately 35,800 km over the surface of the earth, which corresponds to a distance of 42,170 km from the center of the earth, since the earth's radius is 6370 km. Figure 10.1 shows the geostationary orbit configurations. A satellite in geostationary orbit looks at the earth with a total field of view angle of 17.3°. At least three geostationary satellites are required for global coverage. If three spacecraft are positioned symmetrically 120° apart, global coverage can be achieved on the earth's surface with some overlap. The extent of the overlap on the earth's surface will be 4745 km, corresponding to an angle of 42.7° from the center of the earth.

Geosynchronous satellite orbit can be defined as the spacecraft is moved away from the fixed point in the equatorial plane, for example, to 50° inclination with a 24-hour orbital period. The geosynchronous orbit offers high-latitude and daily coverage with a sizable increase in sensor capability over the geostationary orbit. Figure 10.2 shows the northern hemisphere ground trace for this GEO, the trace in the southern hemisphere being a mirror image of that in the north. It is clear that a near-nadir viewing angle of the high-latitude area is available for about several

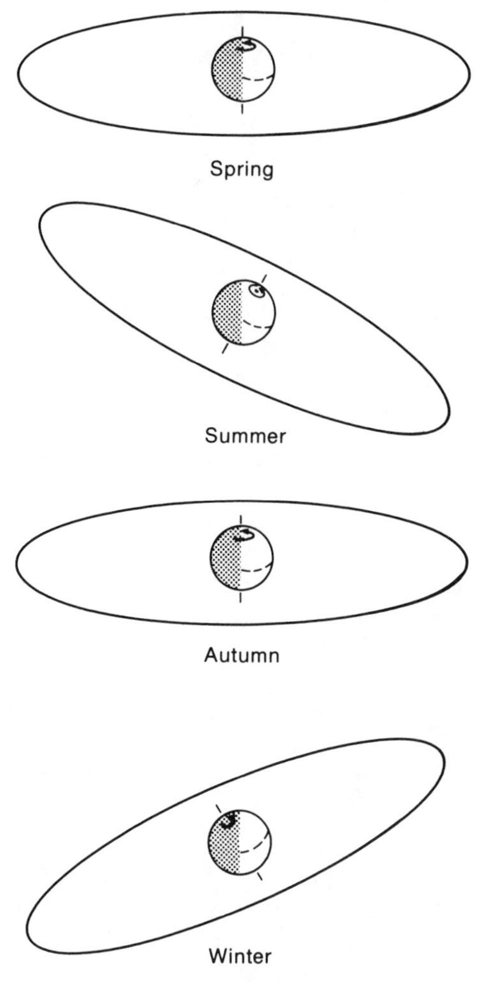

Fig. 10.1 Geostationary orbit configuration.

hours each day. This figure-eight orbit can be a very useful one for the space application of synthetic aperture radar (SAR), because the SAR sensor needs a moving space platform to generate the synthetic aperture for image processing.

Geostationary orbit is the one most often used in earth–space applications because it has certain advantages over geosynchronous orbit. It is convenient for satellite tracking and data processing. Of course, geostationary orbit is useless for earth observation in high-latitude and polar

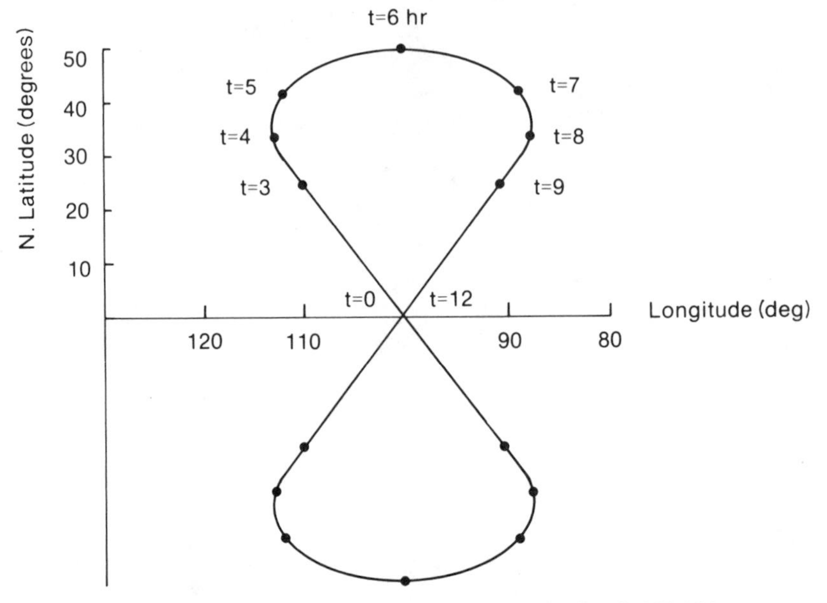

Fig. 10.2 Synchronous figure-eight coverage for a 50° orbit of period 23.98 hr.

Table 10.1

LEO and GEO Comparison

Parameters	LEO	GEO
Orbit type	Sun-synchronous	Geostationary
Orbit height	250–1000 km	35,800 km
Frequency of local area observation	Twice per day (daily)	48 times per day (hourly)
Area coverage	0–90° latitude	70° latitude (maximum)

regions. Table 10.1 shows a comparison of the characteristics of low earth orbit (LEO) and GEO.

10.2 GEO Launch Vehicles

Unmanned expendable launch vehicles such as Delta and Atlas/Centaur have been used to launch GEO payloads to synchronous orbit. The

Table 10.2

GEO Payload Delivery Capability

Launch vehicle	Spacecraft weight at GEO (kg)
Delta	500
Atlas/Centaur	1900
Space Shuttle	6300

Shuttle provides an opportunity for continuous growth in size of the satellite until the maximum length and diameter are reached. Since the payload can be placed by the Shuttle in a LEO at the initial stage, the maximum size capability is also determined by the total size, including the second expendable propulsion stage used to boost the spacecraft into a synchronous orbit. Table 10.2 lists the payload delivery capabilities of the different launch vehicles.

When the Shuttle launch vehicle is used to place spacecraft in geostationary orbit, the following procedures will take place. The Shuttle is launched with the three orbiter Space Shuttle main engines (SSMEs) burning in parallel with the two solid rocket boosters (SRBs). After approximately 2 min, the solid rocket booster propellants are depleted and the SRBs are staged off to be recovered and returned to the launch site. The orbiter ascent is continued by using the three SSMEs, which provide thrust vector control until main engine cutoff (MECO) conditions ensure safe disposal of the external tank (ET). The ET is separated immediately after MECO, and the orbital maneuvering system (OMS) engines provide the additional velocity needed to insert the orbiter into an orbit having a minimum apogee of 250 km. From this low orbit the spacecraft is launched to synchronous orbit by the following operations.

(a) Space Shuttle–spinning solid upper stage (SSUS) operations: One of the Shuttle's capabilities is the injection of Delta-class payloads from standard Shuttle orbits by use of the spinning solid upper stage. The SSUS is a simple, low-cost propulsion stage that obtains its directional stability gyroscopically through spinning. Spin-up of the SSUS and its spacecraft payload is done mechanically prior to launch from the orbiter. The SSUS-D system is mounted vertically and is small enough that as many as four can fit in the cargo bay on one Shuttle flight, resulting in substantial cost savings per spacecraft mission compared to expendable Delta launches. The SSUS-D is also called a payload assist module–Delta (PAM-D). The SSUS-A is an upper stage developed for the Atlas launch vehicles. It is larger than the SSUS-D and must be stowed horizontally in the Shuttle. Prior to deployment, the SSUS-A is erected to a vertical position so that it can be deployed in the same manner as SSUS-D. Figure 10.3 shows a typical trajectory for the transition from low earth orbit to geostationary orbit. After launch to the standard circular orbit, the SSUS and its spacecraft are checked out, pointed in the proper direction by the orbiter, spun up to the required spin rate, and separated from the orbiter

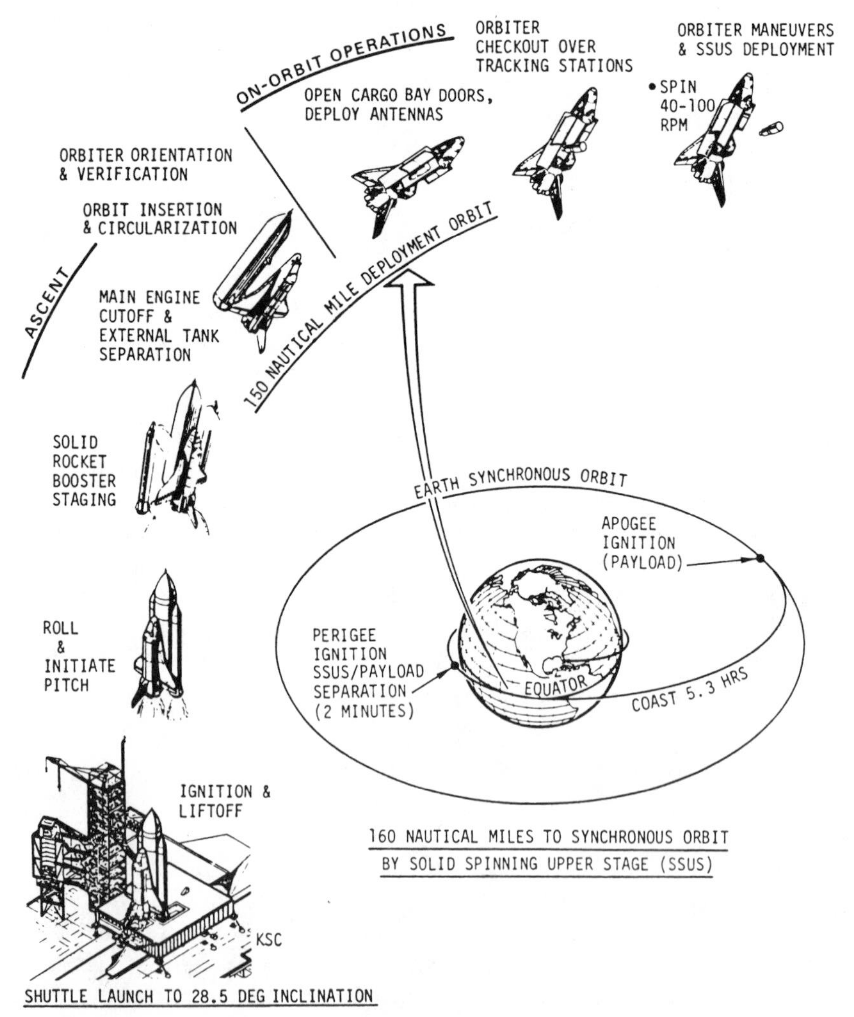

Fig. 10.3. Shuttle GEO launch sequence. (Courtesy of Rockwell International and NASA.)

by means of springs. The Shuttle then performs minor maneuvers and the SSUS perigee kick motor is fired at the proper time to inject the spacecraft into synchronous orbit. The expended SSUS propulsive stage is then separated so that the spacecraft system can continue flight. Upon arrival at apogee of the transfer orbit, an apogee kick motor (AKM) in the spacecraft is fired to circularize the orbit and make course corrections based on the ground tracking results.

(b) Space Shuttle–inertial upper stage (IUS) operation: The inertial upper stage serves as an upper stage of the Space Shuttle system, boosting payloads from Shuttle low earth orbits to higher operational orbits. The IUS comes in at least two versions: a two-stage and a three-stage system. The IUS is a solid rocket upper stage vehicle designed to deliver payloads of different weights to a variety of higher orbits. This is accomplished by using various standard motor combinations. The two-stage IUS has the capability of boosting 2270 kg (and the three-stage IUS, 5450 kg) to geosynchronous orbit. Although it is anticipated that the majority of the IUS missions will be to geostationary orbit, the IUS will also have the capability of delivering heavy payloads to planetary orbits.

10.3 GEO Free-Flyer Satellite Systems

10.3.1 *Applications Technology Satellite (ATS)*

The ATS was designed to test new space sensors and techniques for future environmental satellites from geostationary orbit 35,800 km above the surface of the earth in the late 1960s. The ATS-1 carried a spin-scan cloud camera for synchronous earth observation from space. Longitude scan from east to west was automatically provided by the spin of the spacecraft, while north-to-south scan was produced by using a mechanical step motor to rotate the telescope system and focal plane system. The ATS-3 carried a color spin-scan cloud camera that consisted of a telescope, three optical filters, three photomultiplier tubes (PMTs), and other subsystems for space remote sensing applications from geostationary orbit. The ATS was the beginning of satellite systems useful for earth observations from synchronous orbit.

10.3.2 *Geostationary Operational Environmental Satellite (GOES)*

NASA developed the synchronous meteorological satellite (SMS) for experimental operation and launched the first in 1974 and the second in

1975. These SMS spacecraft systems were followed by the geostationary operational environmental satellites, which make day and night observations of atmospheric vertical temperature, atmospheric vertical moisture profiles, surface temperature, and cloud movements.

There are two GOES flight systems: GOES East and GOES West. Both spacecraft are in an equatorial orbit located above the equator at a distance of 35,800 km from the surface of the earth. The GOES East orbit is arranged to keep the satellite stationary over 75°W longitude; the GOES West orbit is over 135°W longitude. Figure 10.4 shows the area coverage of the GOES satellites.

The total field of view of the GOES coverage is 20° east–west by 20° north–south, in which the full earth disk is centered. It takes about 20 min to image the full earth disk. Two full earth disk images are provided each hour. The GOES sensor is programmed to start new images at 30-min intervals on the hour and the half-hour. The remaining 10-min intervals between images are used for data transmission and ranging.

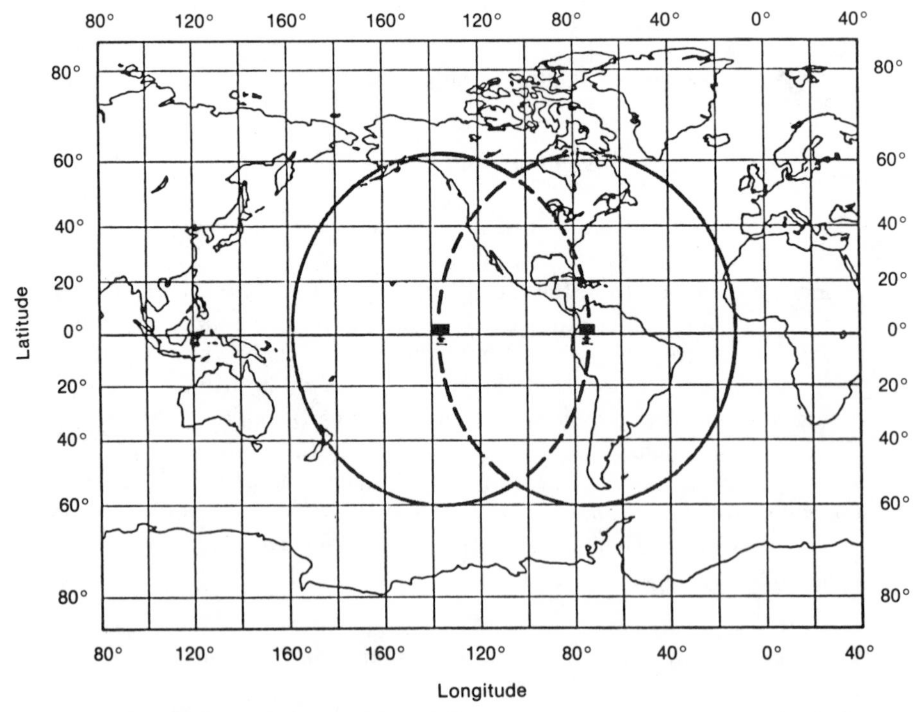

Fig. 10.4. Area coverage of GOES satellite. The solid curves show the area of useful cloud information. The dashed lines show where coverage overlaps.

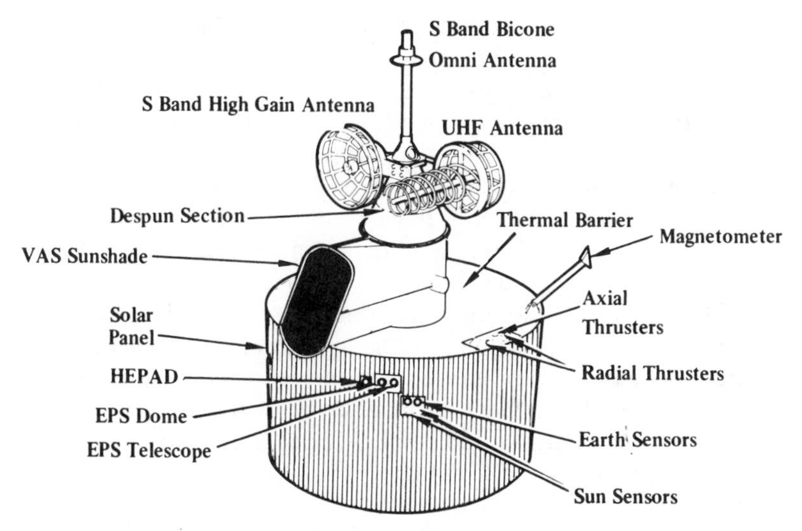

Fig. 10.5 GOES spacecraft configuration. (Courtesy of Hughes Aircraft and NASA.)

The sensor on the GOES 1 to GOES 3 spacecraft is a visible and infrared spin-scan radiometer (VISSR). The GOES 4 and the follow-up spacecraft carry a VISSR atmospheric sounder (VAS). The VAS is used for both imaging and sounding applications, whereas the VISSR is designed for imaging only.

The GOES satellite consists of a spinning section and a despun, earth-oriented antenna assembly. The spacecraft length is 4.4 m and the outside diameter is 2.1 m. Spacecraft spin motion at a rate of 100 rpm provides a simple means for gyroscopically stabilizing attitude and controlling orientation and for maintaining thermal stability. The overall size and envelope of the GOES satellite are compatible with the Shuttle and Delta launch vehicles. Spacecraft on-orbit weight is 347 kg. Figure 10.5 shows the GOES spacecraft configuration.

10.4 Deployable Geosynchronous Platform

The Space Shuttle opens up new opportunities in terms of its ability to deliver large satellites with more applications to geosynchronous orbit. The Shuttle can place up to about 6300 kg of payload into a geostationary orbit. A GEO large satellite system is a large space system in geostationary orbit delivered by the Shuttle or from a LEO Space Station composed of a large structure and integrated modular system. A large GEO satellite has a longer life, because the system can be serviced by automated devices to replace or resupply payloads.

Very large antennas and large telescopes in GEO satellites would make it possible to achieve higher ground resolution for space remote sensing applications. The large antenna would have to be steered electronically rather than mechanically, and a large phased array antenna would be ideal for this special GEO application. Because of the size of the Shuttle cargo bay, any large antenna would require deployment after the launch. Such a deployable antenna could be from 15 to 100 m in diameter.

The GEO large satellite could be deployed outside the Shuttle cargo bay while in the LEO position. After deployment, the satellite system would be checked out by the ground station over a period of a few days. If any systems or payloads failed to function properly, payload specialists would be available in the Shuttle or LEO Space Station to take corrective action. After checkout, the GEO large satellite would be transferred to the geostationary orbit. In the event that more than 1 g is imposed on the large satellite system during the transfer, part of the satellite and some of the payloads might have to be deployed at the GEO position.

The GEO large satellite system anticipated for the 1990s will be a very useful space system for earth observation applications. The GEO sensors will be powerful tools for studying the diurnal and hourly variations of the clouds, surface radiance and polarization, and the atmopshere in the UV, visible, IR, and microwave spectral regions. The future GEO large satellite system will pave the way for new satellite systems of the 2000s in the same way that the ATS space system of the 1960s paved the way for the present GEO applications.

10.5 GEO Space Station Sensor Systems

In the 2000s, geostationary large satellite systems will include larger space sensors, with assembly of large structures and large-aperture systems at GEO to form a manned GEO Space Station. The GEO system will be serviced by a LEO Space Station or by the Space Shuttle. Earth observation from the GEO position will be very useful for studies of climate, environment, land, ocean, weather, and special targets. Applications of a GEO space system are described in the following sections.

10.5.1 *Climate Space Sensor Systems*

Geostationary orbit is a very useful location for demonstrating the technology for passive and active space sensors. The ability to do this in real time with human interaction makes it possible to develop sensor technology in the most optimum way.

For climate observations, a multifrequency, multibeam, large-aperture imaging microwave radiometer and lidar system would be developed and evaluated in GEO to measure several important parameters simultaneously. These parameters are soil moisture, surface temperature, rain rate, atmospheric trace gas concentration, wind speed, and so on. This mission is needed to develop and demonstrate the technology for future GEO operational space systems for the measurement of many important earth climate parameters.

10.5.2 *Environment Space Sensor Systems*

The GEO environment mission will permit the establishment of lidar sensor characteristics in the Space Station environment with the benefits of manned laboratory servicing. High-energy lidar for measurements of atmospheric species concentrations, temperature profiles, pressure profiles, and cloud physics will provide the technology for environmental monitoring from space. The availability of higher electrical power from the GEO Space Station than from the Shuttle will make it possible to obtain vital information for hourly environmental studies. The demonstration of high-energy lidar from the GEO Space Station is of great importance for earth environmental studies.

10.5.3 *Land Space Sensor Systems*

The space sensors currently used for land observations are limited to highly selected spectral bands from the visible to the microwave region. As our understanding of various physical processes increases, there will be a need to develop sensors with an increasingly wide variety of capabilities such as polarization and different looking angles. These new sensors can be best developed with testing from the GEO Space Station.

One of the major problems in developing land observation sensors is that the physical process to be sensed is only partially understood. Variations in atmospheric absorption as a function of land location, effects of various viewing and solar illumination angles, variations in hourly surface temperature, and cloud effects are of interest for the development of new space sensor systems. The GEO Space Station will provide an ideal facility from which to study these hourly variations with new space sensors. In many cases the solution of difficult land observation problems will become clear when all constraints are brought together in a GEO land observation facility. Manned intervention in the sensor development process in space will be critical to the success of land remote sensing.

10.5.4 *Ocean Space Sensor System*

Ocean sensing from GEO has a great potential for contributing to our understanding of the fundamental behavior of the oceans. The ocean microwave package (OMP) is a multipurpose microwave radar that could be flown aboard the GEO Space Station as a development for future space sensors. The OMP combines the wide-swath interferometric altimeter (WIA) with a directional wave spectrometer (DWS). The DWS has already been flown on airplanes and has proved to be a unique technique for measuring the directional spectrum of ocean surface waves by a nonimaging, low-data-rate method. The WIA is an altimeter with two antennas that are separated in the latitude direction to form a latitude interference pattern. Using this technique, it will be possible to measure latitude topography or sea surface slope as well as the longitudinal slope provided by conventional single-beam altimeters. Use of the WIA will improve the accuracy of determination of the oceanic general circulation. Synthetic aperture radar cannot be used at geostationary orbit except in a geosynchronous figure-eight orbit.

Special GEO low-light-detection sensors can be used to map the distribution of marine bioluminescence from space. The distribution and intensity of bioluminescence are functions of both the abundance of small organisms and the processes that mechanically trigger their bioluminescence.

10.5.5 *Weather Space Sensor System*

A space sensor system in LEO can image the same position on the equator only once every 24 hr in the visible spectral region and once every 12 hr in the infrared channel. There is a need to monitor the dynamic atmospheric clouds and weather system at much higher temporal resolution (every 30 min). In the GEO position, the spacecraft is always above the same point on the equator for geostationary applications. This is an advantage of GEO because the weather moves between the earth's surface and the spacecraft, so that elements of the weather system that vary hourly, such as storms and rain clouds, can be observed from this very useful position. Atmospheric wind fields can be derived by measuring cloud motion in successive GEO satellite images. One of the most important GEO applications will be hourly imaging of tropical and mid-latitude areas, where the atmospheric dynamics and variation are most pronounced. Observation of the earth and its atmosphere from GEO offers many advantages. The GEO space sensors can be upgraded with new technology for better spatial resolution and sensitivity. The sensitivity in the sounding mode can be improved by a correlation technique comprising multiple scanning of each scan line during data acquisition and subsequent signal integration over the multiple scans. Because the atmospheric radiance is weak in some sounding channels, a single line must be scanned a sufficient number of times to achieve an acceptable signal level for the temperature inversion computation. Future advanced sensors can be designed with larger apertures and large focal planes by using the TDI approach and special devices for improved performance in sounding applications. Figure 10.6 shows the earth coverage by GEO space sensors with present and future techniques. Linear arrays of large-field-angle sensors will be ready for application in the 1990s. Wide-field matrix array or area array sensors will be ready for coverage of the whole earth disk in the 2000s.

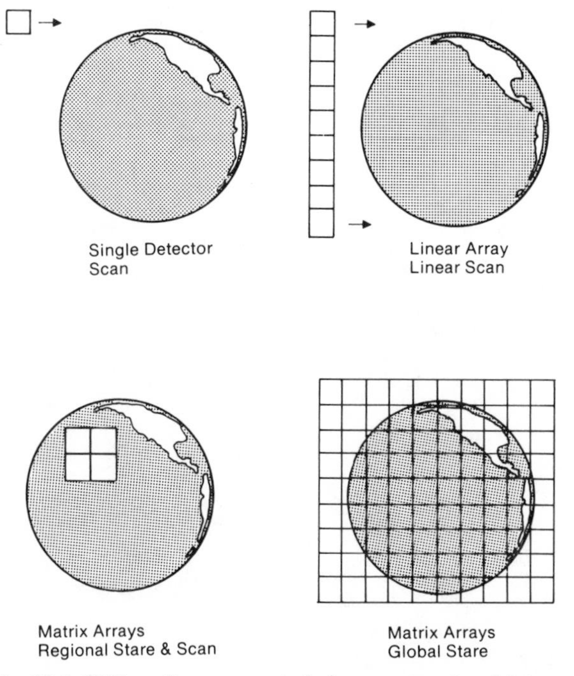

Fig. 10.6 GEO earth coverage techniques, present and future.

General References and Bibliography

Edelson, B. I. (1983). NASA activities in remote sensing. *Proc. Int. Symp. Remote Sens. Environ., 17th,* pp. 7–20.

Fermelia, L. R. (1982). The design and development of GOES. Proceedings of the AIAA 20th Aerospace Sciences. *NASA Conf. Publ.* No. 2227, pp. 35–42.

Hambrick, L. N., and Phillips, D. R. (1980). Earth locating image data of spin-stabilized geosynchronous satellites. *NOAA Tech. Memo.* **NESS-111.**

Heacock, E. L. (1977). Meteorological sensors and related technology for the eighties. *Proc. Int. Symp. Remote Sens. Environ., 11th,* pp. 189–200.

Scofield, R. A., and Gliver, V. J. (1975). The synchronous meteorological satellite (SMS). *Proc. Int. Symp. Remote Sens. Environ., 10th,* pp. 159–162.

Suomi, V. E., and Vonder Haar, T. H. (1969). Geosynchronous meteorological satellite. *J. Spacecr.* **6,** 312–314.

Suomi, V. E., Vonder Harr, T. H., Krauss, R., and Stamm, A. (1971). Possibilities for sounding the atmosphere from a geosynchronous spacecraft. *Space Res.* **11,** 609–617.

Thome, P. G. (1983). Earth remote sensing: 1970–1995. *Proc. Int. Symp. Remote Sens. Environ., 17th,* pp. 35–40.

Yates, H. W. (1970). Results and status of indirect satellite instrumentation development. *In* "Meteorological Observations and Instrumentation," Meteorological Monographs, Vol. 11, pp. 412–432. *Am. Meteorol. Soc.,* Boston, Massachusetts.

Appendix A Space Remote Sensing Parameter Observational Requirements

Parameter	Adequate accuracy	Desired accuracy
Climate:		
Surface albedo	4%	2%
Surface radiation budget	25 W/m^2	10 W/m^2
Solar flux, visible	1%	0.5%
Solar flux, NUV	3%	1%
Solar flux, MUV	10%	5%
Solar flux, FUV	10%	5%
Cloud cover	20%	5%
Cloud-top temperature	4 K	2 K
Cloud albedo	4%	2%
Cloud total liquid water content	50 mg/cm	10 mg/cm
Environment:		
Temperature	2 K	1 K
CO_2	10 ppm	0.5 ppm
CH_4	0.15 ppm	0.05 ppm
N_2O	0.03 ppm	0.01 ppm
H_2O	20%	5%
O_3	0.02 cm	0.005 cm
NO_2	10%	2%
CO	10%	2%
HNO_3	20%	5%
Land:		
Surface albedo	4%	2%
Precipitation	25%	10%
Vegetation cover	10%	5%

Parameter	Adequate accuracy	Desired accuracy
Land: (*cont.*)		
Snow cover	5%	3%
Soil moisture	0.1 g/cm^3	0.05 g/cm^3
Surface temperature	2 K	1 K
Ocean:		
Sea surface temperature	1 K	0.2 K
Evaporation	25%	10%
Sea surface elevation	10 cm	1 cm
Sea surface velocity	10 cm/sec	2 cm/sec
Sea ice cover	5%	3%
Weather:		
Temperature profile	2 K	1 K
Surface pressure	3 mbar	1 mbar
Wind velocity	3 m/sec	1 m/sec
Wind direction	10°	2°
Humidity	30%	7%
Cloud-top height	1 km	0.5 km
Cloud amount	20%	5%

Appendix B | Acronyms

ACS	Altitude control systems
AEM	Atmosphere explorer mission
AFGL	Air Force Geophysics Laboratory
AGU	American Geophysical Union
AIAA	American Institute of Aeronautics and Astronautics
AKM	Apogee kick motor
AMS	American Meteorological Society
APS	American Physical Society
ARC	Ames Research Center
ASPS	Annular suspension pointing system
ATMOS	Atmospheric trace molecular spectrometer
ATS	Applications technology satellite
AVHRR	Advanced very high resolution radiometer
BLIP	Background-limited performance
BNR	Background-to-noise ratio
BUV	Backscattered ultraviolet spectrometer
CCD	Charge-coupled device
CID	Charge-injection device
CLAES	Cryogenic limb array étalon spectrometer
cw	Continuous wave
CZCS	Coastal zone color scanner
DDHS	Digital data-handling subsystem
DIAL	Differential absorption lidar
DMSP	Defense meteorological satellite program
DWS	Directional wave spectrometer
ELV	Expendable launch vehicle
EPA	Environmental Protection Agency
ERBE	Earth radiation budget experiment
ERBS	Earth radiation budget satellite
ERS	ESA remote sensing satellite
ESA	European Space Agency
ESMR	Electrically scanning microwave radiometer

ESSA	Environmental Satellite Service Administration
ET	External tank
ETR	Eastern Test Range
FASCOD	Fast atmospheric signature code
FOV	Field of view
FILE	Feature identification and location experiment
FPA	Focal plane array
FUV	Far ultraviolet
FWS	Filter wedge spectrometer
GARP	Global atmospheric research program
GEO	Geosynchronous earth orbit
GLAS	Goddard Laboratory for Atmospheric Sciences
GMS	Geostationary meteorological satellite
GMT	Greenwich mean time
GOES	Geostationary operational environment satellite
GPS	Global positioning system
GSFC	Goddard Space Flight Center
HALOE	Halogen occultation experiment
HCMM	Heat capacity mapping mission
HDDT	High-density data type
HIRS	High-resolution infrared radiation sounder
HIS	High-resolution interferometer sounder
HRDI	High-resolution Doppler imager
HRE	High-resolution étalon
HRIR	High-resolution infrared radiometer
HRV	High-resolution visible
IEEE	Institute of Electrical and Electronics Engineers
IF	Intermediate frequency
IFOV	Instantaneous field of view
IM	Instrument module
IMC	Image motion compensation
IPS	Instrument pointing system
IR	Infrared
IRIS	Infrared interferometer spectrometer
ISAMS	Improved stratospheric and mesospheric sounder
ITOS	Improved TIROS operational satellite
ITPR	Infrared temperature profile radiometer
IUS	Inertial upper stage
JPL	Jet Propulsion Laboratory
JSC	Johnson Space Center
KSC	Kennedy Space Center
LACATE	Lower atmosphere composition and temperature experiment
LAMMR	Large-antenna multifrequency microwave radiometer
Landsat	Land satellite
LaRC	Langley Research Center
LEO	Low earth orbit
Lidar	Light detection and ranging
LIMS	Limb infrared monitor of the stratosphere
LO	Local oscillator

LOWTRAN	Low-resolution transmittance computer code
LRIR	Limb radiance inversion radiometer
LSI	Large-scale integration
LSS	Large satellite system
LTE	Local thermodynamic equilibrium
LWIR	Long-wave infrared
MAPS	Measurement of air pollution from satellites
MECO	Main engine cutoff
MESSR	Multispectral electric self-scanning radiometer
MFPA	Monolithic focal plane array
MIC	Microwave integrated circuit
MLA	Multispectral linear array
MLS	Microwave limb sounder
MMS	Multimission modular spacecraft
MMU	Manned maneuvering unit
MOMS	Modular optoelectronic multispectral scanner
MOS	Marine observation satellite
MRIR	Medium-resolution infrared radiometer
MSFC	Marshall Space Flight Center
MSI	Multispectral scan imaging
MSS	Multispectral scanner
MSSCC	Multicolor spin-scan cloud camera
MSU	Microwave sounding unit
MTF	Modulation transfer function
MUV	Middle ultraviolet
MUX	Multiplexer
MWIR	Middle-wave infrared
NASA	National Aeronautics and Space Administration
NASDA	National Aeronautics and Space Development Agency
NBS	National Bureau of Standards
NEI	Noise-equivalent irradiance
NEMS	Nimbus E microwave spectrometer
NETD	Noise-equivalent temperature difference
NIR	Near infrared
NOAA	National Oceanic and Atmospheric Administration
NOSS	National Oceanic Satellite System
NRL	Naval Research Laboratory
NSF	National Science Foundation
NUV	Near ultraviolet
OAST	Office of Aeronautics and Space Technology
OCE	Ocean color experiment
OMP	Ocean microwave package
OMS	Orbital maneuvering system
OSA	Optical Society of America
OSS	Office of Space Science
OSTA	Office of Space and Terrestrial Applications
OTV	Orbital transfer vehicle
PAM	Payload assist module
PC	Photoconductive

PIN	*p*-type–intrinsic–*n*-type semiconductor diode
PMR	Pressure-modulated radiometer
PMT	Photomultiplier tube
PPM	Parts per million
PRF	Pulse repetition frequency
PV	Photovoltaic
QE	Quantum efficiency
Radar	Radio detection and ranging
RBV	Return beam vidicon
rf	Radio frequency
RFOV	Resolution field of view
RMS	Remote manipulator system
SAGE	Stratospheric aerosol and gas experiment
SAL	Synthetic aperture lidar
SAM	Stratospheric aerosol measurement
SAMS	Stratospheric and mesospheric sounder
SAR	Synthetic aperture radar
SBRC	Santa Barbara Research Center
SBUV	Solar backscatter ultraviolet spectrometer
SCACU	Spacecraft command and control unit
SCAMS	Scanning microwave spectrometer
SCMR	Surface composition mapping radiometer
SCR	Selective chopper radiometer
Seasat	Sea satellite
SIR	Shuttle imaging radar
SIRS	Satellite infrared spectrometer
SIS	Shuttle imaging spectrometer
SL	Spacelab
SMIRR	Shuttle multispectral infrared radiometer
SMMR	Scanning multichannel microwave radiometer
SMMS	Solar maximum mission satellite
SMS	Synchronous meteorological satellite
S/N	Signal-to-noise ratio
SPAS	Shuttle pallet satellite
SPOT	Satellites probatoires d'observations de la terre
SR	Scanning radiometer
SRB	Shuttle rocket booster
SSCC	Spin-scan cloud camera
SSME	Space Shuttle main engine
SSU	Stratospheric sounding unit
SSUS	Spinning solid upper stage
STALO	Stable local oscillator
STP	Surface temperature and pressure
STS	Space Transportation System
SUSIM	Solar UV spectral irradiance monitor
SWIR	Short-wave infrared
TDI	Time delay and integration
TDRSS	Tracking and data relay satellite system
TFOV	Total field of view

THIR	Temperature and humidity infrared radiometer
TIROS	Television and infrared observation satellite
TM	Thematic mapper
TOMS	Total ozone mapping spectrometer
TOPEX	Topography experiment
TOVS	TIROS operational vertical sounder
UARS	Upper atmosphere research satellite
ULE	Ultra-low expansion
UV	Ultraviolet
VAS	VISSR atmospheric sounder
VHRR	Very high resolution radiometer
VHSIC	Very high speed integrated circuit
VISSR	Visible and infrared spin-scan radiometer
VLSI	Very large scale integration
VTPR	Vertical temperature profile radiometer
WIA	Wide-swath interferometric altimeter
WTR	Western Test Range

Index

T

U

V

W

Z